THE RUSSIA HOAX

THE RUSSIA HOAX

THE ILLICIT SCHEME TO CLEAR HILLARY CLINTON AND FRAME DONALD TRUMP

GREGG JARRETT

An Imprint of HarperCollins *Publishers*

A hardcover edition of this book was published in 2018 by Broadside Books.

HarperCollins books may be purchased for educational, business, or sales promotional use. For information, please email the Special Markets Department at SPsales@harpercollins.com.

FIRST BROADSIDE PAPERBACK PUBLISHED 2019

Library of Congress Cataloging-in-Publication Data has been applied for.

ISBN 978-0-06-287273-9

19 20 21 22 23 LSC 10 9 8 7 6 5 4

TO CATE, GRACE AND LIV,

FOR YOUR LOVE AND SUPPORT

The greatest dangers to liberty lurk in insidious encroachment by men of zeal.

—Justice Louis D. Brandeis (*Olmstead v. U.S.*, 1928, 277 U.S. 438, 479)

CONTENTS

PREFACE

AN INSIDIOUS ABUSE OF POWER

In the spring of 2018, James Comey began peddling his vainglorious book with an imposturous title, *A Higher Loyalty*. It reads like a Harlequin romance, except that the protagonist is in love with *himself*.

Comey sermonizes about lies and lying people. This is perversely ironic coming from a man who, more than anyone else, is responsible for the most notorious hoax in modern American history.

As director of the Federal Bureau of Investigation, Comey launched a dilating investigation into Donald J. Trump in the summer of 2016. There was not a whiff of credible evidence to legally justify the probe. So, in a deception worthy of a solid street hustle, Comey labeled it a "counterintelligence matter." The clever feint allowed for a covert criminal investigation in search of a crime, reversing and bastardizing the legal process.

This is what abuse of power looks like.

For the better part of two years, the FBI, U.S. intelligence apparatus, mainstream media, Democrats, and, eventually, Special Counsel Robert Mueller, and his team of partisans peered furiously into every obscure corner and crevice for some proof that Trump "colluded" with Russia to steal the 2016 presidential election. They seemed to believe that Trump could not possibly have won absent some treasonous conspiracy hatched by him in the bowels of the Kremlin. They were convinced he was an illegitimate president. Besides, Comey and his minions didn't like Trump—the man or his politics. They loathed him. The vaunted director, who surely knew better than naïve and gullible voters, would be the savior of the nation.

Armed with immense power, Comey and other top officials at the FBI became the law unto themselves. The end would justify the means, if only collaboration with the Russians could be found. Failing that, the investigation itself could be manipulated to create the illusion of wrongdoing. The ensuing scandal would be sufficient to annul Trump. It wasn't merely that the flamboyant businessman-turned-candidate would upset the natural order of the political landscape, but as president he would likely show the director the door at the Hoover Building in Washington, D.C. There is nothing that will motivate a man like Comey more than the prospect of power forfeited. Something had to be done.

But first, Comey and his confederates had to turn an even tougher trick by exculpating Hillary Clinton, the Democratic presidential nominee, from the sundry crimes she appeared to have committed by storing copious classified documents on her unauthorized private computer system at the Clinton homestead. Despite a subpoena to preserve her records, tens of thousands of government documents were deleted, her server wiped clean, and numerous devices destroyed. How exactly could the FBI circumvent the inevitable multicount indictments against Clinton for everything from obstruction of justice to mishandling of classified documents?

It would not be easy. Criminal codes would have to be creatively reinterpreted or deliberately misconstrued. A grand jury must be avoided or diminished at all costs. But the Department of Justice would acquiesce since it was brimming with Clinton allies, especially Attorney General Loretta Lynch. Best of all, President Barack Obama had sent out the clarion call on national television that his chosen Democratic successor should not be prosecuted since she never "intended" to jeopardize national security, notwithstanding what he characterized as her "carelessness."

Comey was instrumental to the scheme. He penned an "exoneration statement" well before the FBI interviewed seventeen key witnesses, including Clinton herself. Then he commandeered the authority of the attorney general in delivering his infamous pronouncement that no charges should be brought. No mention was ever made of another vexing FBI probe into whether Clinton had engaged in corrupt acts with the Russians by hanging the equivalent of a "for sale" sign on her office door as secretary of state. That investigation quietly vanished under the auspices of the Obama administration. Thus, there was never a serious examination of Clinton and the various laws she seemed to have plainly violated. The entire exercise was a preconceived charade with a predetermined outcome.

Evidence of the suspected scheme to protect Clinton from any criminal prosecution was divulged in a five-hundred-page report by the Justice Department's inspector general, Michael Horowitz, that was made public two weeks after *The Russia Hoax* went to print. In one of the more stunning revelations, Comey claimed he had no memory of the moment he decided—and put down in writing—that Clinton committed crimes. He admitted composing a statement on May 2, 2016 concluding that Clinton was "grossly negligent." Those pivotal words have a distinct legal meaning and are drawn directly from a federal statute that makes it a felony to handle classified documents in a "grossly negligent" manner.

Comey used the exact phrase not once, but twice. Based on his finding, Clinton should have faced a multiple-count indictment since the FBI discovered that she had stored 110 classified emails on her unauthorized, private computer server. Other people had been prosecuted for similar conduct that jeopardized national security in violation of the law. Yet Comey—despite characterizing Clinton's actions with the clear language denoting violations of law–saw to it that no charges were ever brought against Clinton.

Under questioning, Comey admitted to the I-G that he authored the May 2 statement and penned every word of it himself. But then he offered the implausible claim that "he did not recall that his original draft used the term 'gross negligence,' and did not recall discussions about that issue."

Comey's amnesia is preposterous. He would have us believe that, as FBI director, he memorialized in print his decision that the leading candidate for president of the United States had committed crimes, yet later could not recollect anything about the most important decision of his career.

The truth is that Comey surely remembers what he wrote, because he participated in subsequent discussions with top officials at the FBI about Clinton's "gross negligence." Several meetings were held on the subject and contemporaneous notes prove that Comey was in attendance. Those records show that although Comey seemed convinced that Clinton was "grossly negligent" in violation of the law, he was determined to clear her notwithstanding. To achieve this somersault and absolve the soon-to-be-

Democratic nominee, the legally damning terminology would have to be stricken from his statement.

Metadata shows that on June 6, the FBI's lead investigator on the case, Peter Strzok, sat down at his office computer to cleanse his boss's statement of the vexing term, "gross negligence." With the help of his paramour and FBI lawyer Lisa Page, the words "extremely careless" were substituted to make Clinton appear less criminally culpable. When questioned herself, Page told the I-G that "to use a term that actually has a legal definition would be confusing."

It most certainly would. After all, how could Clinton be exonerated under the "gross negligence" law if that very phrase was used to describe her behavior? The two phrases actually mean the same thing in the law, but only one appears in the statute.

Strzok and Page also expunged from Comey's statement his reference to another statute that Clinton had plainly violated. She should have been charged under the statute's "intent" provisions. With Comey's consent and encouragement, the pair sanitized his findings of fact and contorted his conclusions of law. Clinton, who had not even been interviewed by the FBI yet, was free and clear. The investigation was a sham.

While Comey claimed he could not remember writing the words that should have indicted Clinton, he had complete recall of the law he either misinterpreted or misunderstood. He told the I-G that he thought "Congress intended for there to be some level of willfulness present even to prove a 'gross negligence' violation." Yet no such "intent" is required. If Comey had ever read the legislative history, he would have known that in 1948, Congress amended the original Espionage Act of 1917 to add a new "gross negligence" provision that did not require intent or willfulness.

Amnesia must be contagious at the FBI. Testifying before Congress, Strzok feigned no recollection of using his computer to make the critical alteration that cleared Clinton. He did, however, directly implicate the FBI director. "Ultimately, he [Comey] made the decision to change that wording," said Strzok.

Text messages exchanged between Strzok and Page provide a powerful narrative of how severe bias against Trump drove top officials to clear Clinton and target Trump in an effort to prevent him from becoming president. The most damning exchange discovered by the I-G occurred on August 8, 2016, one week after Strzok signed formal documents launching the FBI's investigation of Trump.

Page asked, "[Trump's] not ever going to become president, right? Right?" Strzok responded, "No. No, he won't. We'll stop it." When questioned by Congress, Strzok once again pleaded amnesia insisting that he did not recall writing the infamous text. Yet he then testified, "What I can tell you is that text in no way suggested that I or the FBI would take any action to influence the candidacy."

That is a remarkably dexterous explanation for something he does not remember doing. When confronted with myriad other messages extoling Clinton and disparaging Trump, Strzok had the temerity to say, "I do not have bias." He added, "Those text messages are not indicative of bias." No one with an ounce of intelligence could possibly buy the self-serving rubbish that Strzok was peddling. This included the inspector general who, after an exhaustive investigation, concluded that the Strzok-Page communications "are not only indicative of a biased state of mind but imply a willingness to take official action to impact a presidential candidate's electoral process."

This pervasive anti-Trump bias propelled top officials at the FBI to pursue its imaginary case of "collusion" against Trump. The evidence was scant, even nonexistent, until a scurrilous anti-Trump "dossier" surfaced. It was secretly funded by the Clinton campaign and delivered to the FBI by a former British spy who *claimed* to have spoken to unidentified Russians. The FBI first met with the author of the "dossier" on the same day that Comey absolved Clinton. But try as it may, the bureau seemed unable to verify its inflammatory contents alleging that Trump had been collaborating with Russia for at least five years. No matter. The FBI appropriated the document as a pretext for spying on a Trump campaign associate while simultaneously advancing their investigation of the candidate who unexpectedly became president. They actively hid the fact that Clinton and Democrats had underwritten the specious document. At the same time, Comey's FBI deployed an undercover government informant to infiltrate the Trump campaign to induce (or entrap) associates into saying something incriminating.

The unverified "dossier" was leaked to the press and, by the time Trump was sworn in as president, the mainstream media had its "collusion" tale. It was off to the races as journalists propagated their nonstop narrative that Trump was undoubtedly guilty of collaborating with Moscow to win the election. But a curious thing happened on the way to their twin goal of indictment and impeachment. "Collusion" is not a crime, except in antitrust law. Reporters and anchors never bothered to cite a specific statute because none could be found in the criminal codes. Not that they ever bothered to look. They were satisfied, indeed anxious, to level the accusation as a legal conclusion because their bias against Trump was so impassioned and pervasive that it became impervious to the facts.

The media's miscalculation was compounded by the inconvenient absence of any incriminating evidence to prove this invisible offense called "collusion." They were adamant that meeting with, or talking to, a Russian, even on American soil, was somehow an undefined crime unprotected by the First Amendment. Any chance encounter or a casual handshake with a Russian was vilified. Breathless media accounts inflated trivial contacts beyond reason. They fanned tiny embers, repeatedly arguing that this much smoke means there will eventually be fire. When the truth finally sunk in that "collusion" appeared to be a myth, the press seamlessly shifted to a different accusation that Trump must have obstructed justice by attempting to impede the FBI's investigation.

In this, Comey once again became the central character. When he was fired as FBI director for repeated violations of the bureau's regulations, he admitted pilfering government memorandums, which he improbably claimed as his own, for the sole purpose of triggering a special counsel who just happened to be his longtime friend and ally, Robert Mueller. When Comey testified before Congress, he all but accused the president of attempting to obstruct his investigation, even though the detailed memos offer no such allegation. It was a Machiavellian move to perpetuate the fabricated case against Trump.

Over the course of a year, Mueller's investigation did manage to render several indictments. However, none of them had anything whatsoever to do with "collusion" with the Russians. Did Moscow meddle in the presidential election? Yes, in much the same way it had tried to tamper for decades. Their design was to sow discord in the Western democracy they so detest. But the notion that anyone in the Trump campaign had somehow been complicit was dispelled when the indictment of numerous Russians revealed that no Americans knowingly participated. To date, no known actionable proof of "collusion" involving Trump has surfaced.

They did, however, sow their desired political discord, thanks to a deluded and compliant American left.

Comey devoted several months to a cross-country "adoration tour," hawking his book and pocketing millions of dollars. In the process, he accused Trump of being morally unfit, ignoring that he was all too willing to serve the same president he now so freely condemns. There is something profoundly incongruous about a man who pontificates about professional ethics but seems bereft of them himself. His hypocrisy was underscored when, three days after his book was published, it was confirmed that the inspector general of the Justice Department was investigating Comey for leaking the presidential memos and mishandling classified information. FBI agents were forced to conduct searches to contain the leak. It is conceivable that Comey committed the very offenses he helped Clinton escape. Self-righteous people tend to self-immolate.

The tale of *The Russia Hoax* is a labyrinth of intertwined stories. Clinton's penchant for secrecy surrounding her dealings with foreign governments and money sources almost certainly propelled her to set up her private server. When caught, she spun a series of deceptions to justify her actions. Comey and company abided by distorting the law to clear her path to the presidency and influence its outcome. This same small group of Washington insiders then abused the power of their positions to conjure a case against Trump to defeat him in the election. When that failed, they doubled down to frame the new president for crimes he did not commit. The "dossier," which Comey confessed was "salacious and unverified," was their *in flagrante delicto* or "smoking gun." Its lack of authenticity was irrelevant. It would be politicized and weaponized to achieve the objective of destroying Trump and driving him from office.

The provenance of this book can be found in the many opinion columns I began composing three years ago in which I argued that Clinton mishandled classified documents and violated the law.

When Comey inexplicably absolved her, and launched a fraudulent investigation of Trump, even more columns were written. But the shape and theme of the book did not fully materialize until October 2017, when I published *The Trump-Russian Collusion and Other Great Hoaxes*, followed by a column titled "Did the FBI and the Justice Department Plot to Clear Hillary Clinton, Bring Down Trump?" By this time, it had become evident to me that the FBI, as well as the Department of Justice, had become cesspools of corruption that played a direct hand in attempting to influence the election and subvert the democratic process.

Following the publication of *The Russia Hoax*, Mueller and his assembled team of partisans continued their pursuit of Trump with scant success. The special counsel eventually abandoned his quest to personally interview the president, accepting instead written responses to a limited number of questions. Since Mueller had no legal right to interrogate the president for exercising his constitutional authority, the issue of obstruction of justice was excluded from the questioning. In his answers, Trump repeated the same statements he had made publicly so often over the course of nearly two years—he never "colluded," conspired, or coordinated with Russia to win the 2016 election.

By all appearances, the special counsel investigation appears to be reaching an ignominious end. Not a single charge involving "collusion" with Russia has been brought. There was no secret rendezvous in Moscow, no clandestine bank transfer, no corrupt conveyance, and no incriminating document that would reveal some grand conspiracy to steal the presidency. Trump's critics were absolutely convinced—and promised–that such evidence would be known by now. Given the continuous stream of leaks coming from the special counsel, it is nearly inconceivable that such credible proof of "collusion" will suddenly emerge to justify the endless and unwarranted accusations that have bedeviled the president.

This is not to say that Mueller will resist leveling some stinging criticism when he issues his final report. The special counsel is nothing if not an implacable antagonist. He seems determined to find fault, however unreasoned or remote. Trump's inveterate detractors in Congress and the mainstream media will surely seize upon any minor or casual contact with Russia and inflate it into a treacherous plot. Demands for impeachment will persist, as they have since Trump was barely sworn into office. All the while, Mueller has ignored the egregious conduct of top officials at the FBI and Department of Justice who schemed to clear Clinton and frame Trump.

The greatest peril to democracy today is not a foreign force, but the abuse of power from within. Those who operate under color of authority are prone to exploit it for political reasons and personal gain. In a government of laws, they too often fail to follow the law themselves. As Justice Brandeis put it, "if the government becomes a lawbreaker, it breeds contempt for the law; it invites every man to become a law unto himself; it invites anarchy."

The most celebrated form of governance known to the world is democracy. It is a system of, by, and for the people. Because of this personal construct, democracy is susceptible to the same human frailties that afflict us all—greed, prejudice, hubris, intellectual dishonesty, and moral weakness. Government, therefore, is only as sound and effective as the people who are empowered to run it at any given time. This is its fundamental imperfection.

Sometimes, driven by these flaws of the human condition, democracy can be trifled with by those who hold the reins of power in government. Despite carefully devised checks and balances, individual positions are nevertheless misused in ways unseen by the public. Systems are unduly influenced in defiance of both the law and conscience. Misdeeds are covered up.

This is the story of *The Russia Hoax*. It is a cautionary story of the damage wrought when those who serve in government seek to undermine it.

Comey and other participants in this hoax were not interviewed for the book. I doubted they would consent and distrusted their credibility. My conclusions about their motivations and intent are drawn from the extensive public record of their conduct in these matters and interviews with current and former employees of the FBI and Justice Department about such conduct, and, based on this record, no other conclusion is reasonably supportable.

GREGG JARRETT
NOVEMBER 2018

THE RUSSIA HOAX

CHAPTER 1

HILLARY CLINTON'S EMAIL SERVER

Convenience is not a legal principle.

—Lord Justice Edward Pearce, Queens Bench, High Court of Justice of England and Wales (1961)

This is a story of corruption. It begins, as it must, with Hillary Clinton.

Clinton's determination to set up a private email server in the basement of her home to use as the exclusive method for electronically communicating all of her official business as secretary of state was a fateful decision. At its core, the idea was fundamentally reckless. It also appears to have been criminal. More than any other event or factor, it led to the inexorable end to her aspiration to be President of the United States.

THE EMAIL SETUP

Unraveling Clinton's private server apparatus was complicated by the hidden nature of its design. The domain was not registered in her name, but under a separate identity even though the IP address was connected to the Chappaqua, New York, residence of Bill and Hillary Clinton. Indeed, the server itself was installed in the basement of their house.[1] The State Department never certified the server as secure, as its rules require.[2] Thus, anyone within the home, including those without security clearance, had physical access to it, as did the two individuals who were instrumental in its installation, operation, and maintenance, Justin Cooper and Bryan Pagliano. Neither was cleared for handling classified information.[3]

Why would Clinton want to keep all her communications as secretary of state on a private server? The obvious answer is often the correct one. With no emails on a government account, Clinton would be able to avoid complying with various requests filed by the media and the public under the Freedom of Information Act (FOIA). This explanation is reinforced by the conscious decision she and her staff made not to utilize an electronic program called SMART, which would have preserved her records for the government.[4]

The motivation behind Clinton's decision to use a secret server invites the question: if someone has nothing to hide, why hide? People who engage in improper, if not illegal, activities often cloak them in darkness. They seek to obscure their illicit behavior. Clinton may have wanted to hide decisions and mistakes that would look bad when she ran for president. She may have wanted to keep from public view any evidence that she and her husband used her position of power in government to enrich themselves. As will be detailed in Chapter 4, their charity, the Clinton Foundation, served as the conduit for hundreds of millions of dollars that flowed from foreign sources at the same time Hillary appeared to have exerted influence on their behalf. At the same time, Bill pocketed tens of millions for speaking engagements overseas, many of them connected to his wife's work.

The improper use of Clinton's surreptitious server stands in stark contrast to the promises she made before taking office. While campaigning for president in 2008, Clinton had promised complete transparency and vowed to fulfill all FOIA requests.[5] In retrospect, her public assurances were nothing more than a carefully crafted deception. Her private server afforded her a clever and clandestine way to evade the public disclosure of her communications and to hide her self-serving deals.

On or about the day Clinton was sworn in to office, January 21, 2009, the State Department activated a classified email account on its secure government server for her benefit. Clinton never used it.[6] Instead, she chose to utilize her unsecured private server for all her work-related communications, including classified documents.

How this went undetected during her four-year term remains inexplicable, unless those with knowledge were complicit in enabling her scheme or turned a blind eye to Clinton's conduct. Staffers at the State Department and the White House, including President Obama, regularly communicated with the secretary of state at her private email address.[7]

Documents uncovered by the FBI show that Obama used a pseudonym in communicating with Clinton on her private email account.[8] Sometimes he did so while his secretary of state was overseas using an unprotected mobile device. When the FBI showed a copy of one such Obama email to Clinton's top aide, Huma Abedin, she exclaimed, "How is that not classified?"[9] Indeed, it surely was classified. Thereafter, the White House refused to disclose the contents of any of the president's emails involving his fake name. Obama was not only concealing his communications with an alias, he was mishandling classified information in the same negligent manner as Clinton. This may explain why the FBI and the DOJ were motivated not to charge Clinton. If they did so, Obama's mishandling of classified communications would be exposed.

It is hard to imagine that the president and others never noticed that Clinton was using a nongovernment, nonsecure private account. Indeed, emails prove that many of them did know.[10] But the secretary of state was surrounded by long-time Clinton cronies. They shielded her. They were not about to challenge the person they were certain would become the next president of the United States.

Clinton's disdain for regulations and statutory law was also evident in her use of BlackBerrys linked to her home server, despite warnings from State Department security personnel.[11] According to an FBI report, she insisted on keeping her mobile device in an area where they were not allowed because hackers can infiltrate them to record classified discussions.[12] While on foreign visits, Clinton incessantly used her unprotected BlackBerry to send or receive dozens of confidential emails that were susceptible to interception by foreign governments.[13] She ignored security protocols and the law with impunity.

CLINTON'S SERVER DISCOVERED

Before Clinton left office in early 2013, FOIA requests seeking information about her emails resulted in terse statements from the State Department that no such records could be located.[14] No wonder. Clinton had stealthily kept all her emails on her private server and never preserved them on a government account as regulations required. These vacuous responses to FOIA demands should have immediately raised red flags of alarm.

It was not until 2014 when the House Benghazi Committee began investigating the September 11, 2012, terrorist attack in Libya during Clinton's tenure that the true nature and extent of her secret server began to slowly unfold. During the summer of 2014, the Committee demanded access to her emails which prompted the State Department to contact Clinton who was now out of office. Negotiations among lawyers ensued and, several months later on December 5, 2014, Clinton reluctantly handed over 30,490 emails from her private account. Simultaneously, she withheld some 31,839 emails, insisting they were private and personal.[15]

At least one of Clinton's lawyers, Heather Samuelson, who helped sort the emails that included classified documents, appeared to have no security clearance in violation of the law.[16] Although Clinton later stated under oath that she had provided all work-related emails, this was not true. According to then-FBI director James B. Comey, thousands of work-related emails were improperly held back by Clinton.[17]

Upon examining the large cache of documents provided by Clinton's attorneys, the Benghazi Committee realized something was amiss: as secretary of state, Clinton had never used her secure government account for any of her communications. Instead, she had conducted all of her business on an unsecured account on a private server located in the basement of her residence. It was, in essence, a hacker's paradise. Confidential information was exposed to theft. Foreign governments with their sophisticated computer expertise could readily access America's classified secrets. Clinton's egregious breach of rules, regulations, and laws jeopardized national security.

New York Times broke the story wide open on March 2, 2015, creating a firestorm of controversy with immense political consequences for Clinton, just as she prepared to launch another bid for president of the United States in the next month.[18] Her actions, decried by many as irresponsible and reckless, called into question both her judgment and competency to hold the highest public office in the land. More important, Clinton's unmitigated disregard for the law quickly metastasized into allegations of criminal wrongdoing.

CLINTON'S ATTEMPTS TO COVER UP

Initially, Clinton employed a phalanx strategy of "no comment." It did not quell the gathering storm. On March 10, 2015, she decided to address the matter with reporters by holding a news conference.[19] It was an unqualified disaster. Under questioning, she appeared unsteady in front of cameras, unsure of her answers and seeming to prevaricate at every turn. Her statements stretched credulity.

Clinton insisted, "I did not email any classified material to anyone on my email. There is no classified material."[20] That remark was brazenly untrue. Clinton had sent numerous classified emails on her unauthorized and unsecured server.

Later, when faced with evidence that contradicted her claim, Clinton changed her story significantly by stating, "I am confident that I never sent or received any information that was classified *at the time* it was sent or received."[21] In other words, she claimed the classifications were made retroactively. But this amended statement was also false. In truth, the FBI confirmed that 110 of her emails contained classified information *at the time they were sent or received*.[22]

As the emerging facts continued to belie Clinton's statements, she altered her story for a third time by claiming, "I never sent or received any email that was *marked* classified."[23] Translation: the classified documents on her email server were unmarked and, as such, she could not be expected to know that they were classified. But this was another untrue statement. It earned her four Pinocchios, the worst rating, from the *Washington Post* "Fact Checker," Glenn Kessler.[24]

Many of Clinton's emails were, indeed, marked classified, as the FBI later revealed. While testifying before the House Committee on Oversight and Government Reform, Director Comey called Clinton's statement "not true."[25] Moreover, the markings were irrelevant under the law, since the content—not the marking—made them classified.

It is important to note that *knowledge* of a document's classification status is of no legal consequence. Thus, it was no defense for Clinton to claim she did not *know* that matters were classified. As secretary of state, she was privy to many of the nation's most guarded secrets. Did they have to be specifically marked "Top-secret" or "classified" for her to recognize them for what they were? Over the course of four years, she was America's top diplomat with direct access to the most sensitive information and data. Assuming a modicum of competency, classified information should have been recognized by her instantaneously.

Still, Clinton seemed to be arguing her own incompetence. That is, she should not be held legally liable because she was too uninformed or inept to recognize classified materials without their markings. But ignorance and maladroitness are not defenses under the law. This rendered Clinton's explanations of her actions all the more unconvincing.

One of the more outlandish, if not comical, explanations offered by Clinton during her FBI interview was that she did not realize that the parenthetical "C" meant classified material at the confidential level when it appeared on documents.[26] She claimed that she thought it simply a way of organizing messages in alphabetical order, although there were no other letters used in this way on the same records.

Several former FBI officials interviewed for this book accused Clinton of deliberately deceiving the FBI during her interview. Bill Gavin, who first joined the bureau in 1967 and became assistant director, did not believe her rationale for one moment:

> "C" stands for cunning. She knew exactly. You cannot have access to classified information without going through the briefings, fully understand what the briefings are telling you and then signing off on documents admitting that you understood what the briefings have said. To say she didn't understand what "C" meant is just a bald-faced lie. She understood what was going on. She didn't care because she knew she was allowed to do anything she wanted to do.[27]

Oliver "Buck" Revell, who spent thirty years with the FBI and rose to become the associate deputy director, called Clinton's alibi absurd:

> She didn't know what the "C" on top of the documents meant? No one believed her. She didn't even believe it herself. That's so nonsensical. It's just ridiculous. I think it reflected a total disregard of the system and the rules that have been established to protect the system. She exposed the nation's secrets through her emails. She took the privileged approach to virtually everything she's ever done in her life,

> disregarding the rules and regulations that apply to everyone else.[28]

All the excuses offered by Clinton were subsequently contradicted by both the FBI and the inspector general at the State Department. Undaunted by her series of demonstrably false statements, she then resorted to a blanket assertion that her actions were legally permissible. In a July 7, 2015, interview, Clinton insisted:

> Everything I did was permitted. There was no law. There was no regulation. There was nothing that did not give me the full authority to decide how I was going to communicate. Everything I did was permitted by law and regulation.[29]

This was another audaciously untrue assertion by Clinton that earned her three Pinocchios from the *Washington Post* "Fact Checker."[30] Her claim was also disproven by the State Department's own inspector general who took office after Clinton left her post but investigated her conduct while she was in office. First, the IG found that Clinton "did not comply with the Department's policies that were implemented in accordance with the Federal Records Act."[31] Second, the IG determined that Clinton never sought approval for her server and, had she done so, it would have been rejected as a risk to national security.[32] Third, the IG learned that employees who voiced concerns about the private server were instructed "to never speak of the Secretary's personal email system again."[33] Fourth, the IG concluded that any classified material "never should have been transmitted via an unclassified personal system."[34]

A federal judge also voiced his disbelief in Clinton's rationalizations and alibis. During a FOIA lawsuit by a nonprofit group called Judicial Watch to recover emails, U.S. District Judge Emmet G.

Sullivan scolded Clinton over her disdain for the law when he said, "We wouldn't be here today if this employee had followed government policy."[35]

CLINTON KNEW SHE BROKE THE LAW

Clinton certainly knew that she was breaking the law when she made the conscious decision to use a personal server for all her work that would, as secretary of state, necessarily involve classified information, including "top-secret" records.

On January 22, 2009, the day after she was sworn into office, she signed a document titled "Classified Information Nondisclosure Agreement," which stated the following:

> I have been advised that the unauthorized disclosure, unauthorized retention, or negligent handling of classified information by me could cause damage or irreparable injury to the United States or could be used to advantage by a foreign nation. I have been advised that any unauthorized disclosure of classified information by me may constitute a violation or violations, of U.S. criminal laws.[36]

Importantly, the document instructed Clinton that classified material may not always be marked as classified on the face of the documents, but that it would still be considered classified information. Thus, she was bound by law to protect it in a secure place, not in her home.

A second document entitled "Sensitive Compartmented Information Nondisclosure Agreement" executed by Clinton that same day acknowledged that she had received a comprehensive security briefing that instructed her on all the applicable laws involving sensitive information and classified materials and how those documents must be kept secure.[37]

Clinton could hardly claim amnesia as a defense, since she sent an order to every State Department employee two years later, on June 28, 2011, cautioning them not to do what she was doing covertly in violating the law. Clinton's order warned: *Avoid conducting official Department business from your personal email accounts.*[38]

There is little doubt that Hillary Clinton flagrantly and shamelessly ignored not only State Department regulations, but federal laws making it a crime to mishandle classified documents. She may also have had a direct hand in the destruction of evidence.

On March 4, 2015, Congress sent a letter to Clinton instructing her to preserve all her emails.[39] Subpoenas were sent immediately thereafter. The request included both work-related emails and any of those she had previously withheld as "personal."

Three weeks later, on March 25, more than thirty thousand of Clinton's emails were deleted. It is unclear how many of those emails might have been relevant to the subpoenas since they have vanished. Her server was wiped clean by using a file-deleting software program called BleachBit.[40]

The act of defying a congressional subpoena is impudent at best, criminal at worst. Destruction of records in any federal investigation constitutes obstruction of justice and a violation of the Federal Records Act.[41]

While the FBI did manage to recover several thousand deleted emails that were, in fact, work-related, no charges were ever brought against Clinton or anyone else involved in their deletion and destruction.

Former assistant director of the FBI Steve Pomerantz is convinced Clinton knew she was breaking the law, but didn't care:

> It is consistent with everything I know about the Clintons. They make their own rules, and it's wrong. Hillary Clinton engaged in conduct that was dangerous to the national security of the United States. And, of course, lying about it only compounds the problem. The Clintons have a history of lying. That's what they do. First they commit the offense, then they lie about it. That's what they do.[42]

It is revealing that Clinton spent five months refusing to turn over her email server to either the Benghazi Committee or the FBI, despite repeated demands to do so. In the end, she relented when faced with the likelihood that Bureau agents would begin serving search warrants to seize it on their own. Fearing a knock on the door of her private residence, Clinton finally capitulated.

Although Clinton is a trained lawyer who graduated from Yale Law School, she seemed oblivious to the law. Or, more likely, she felt that the law did not apply to her. Yet there is no station in life or standing in government that absolves someone from criminal conduct. In this way, we are all creatures of the law and are bound to obey it. An orderly society cannot function if it permits individuals to disregard the law with impunity.

This fundamental principle, enunciated by the U.S. Supreme Court more than a century ago, is what gives sustenance to our democracy. Without it, lawlessness, chaos, and tyranny at the hands of the few would surely ensue. It follows, then, that Clinton is no higher or lower than any American. She must abide by the rule of law regardless of her condition or circumstance.

Former independent counsel Joseph diGenova is convinced Clinton did not care about complying with the law:

> Mrs. Clinton knew precisely what she was doing. She chose to ignore the law. There is simply no explanation other than that. The requirements of the law were too oppressive for her. She was different. She viewed herself differently. These restraints and constraints were something that were for other people—not for her—because she was a special, special person.
>
> I think what is true about Mrs. Clinton is that she had enablers around her all the time. No one was strong enough or willing to tell her she shouldn't do certain things because they all wanted to be part of the big game which ended with her being President of the United States.[43]

THE LAW

Whenever a government official creates or obtains a record during the course and scope of his or her employment, the Federal Records Act and the Foreign Affairs Manual require that it be *captured and preserved by the agency.*[44] Clinton did not do so at any time during her service as secretary of state, thereby violating the law on a daily basis for four years. Every single one of her work-related emails constituted federal records, not her personal property.

It is no excuse under the law that others with whom Clinton communicated via email within the State Department may have kept their own records in the agency system or elsewhere. The records on Clinton's system were never concurrently archived at the

Department of State. She communicated with many other people who were not employees of the agency or government and these emails were never captured or archived. At no time was Clinton ever authorized to create her own private server for record-keeping.

Clinton should have been charged with stealing government documents. Converting government records for personal use is a felony under the law. It is, in a word, theft. Consider 18 U.S.C. 641:

> Whoever embezzles, steals, purloins, or knowingly converts to his use or the use of another, or without authority, sells, conveys or disposes of any record, voucher, money, or thing of value of the United States or of any department or agency thereof . . . or whoever receives, conceals, or retains the same with intent to convert it to his use or gain, knowing it to have been embezzled, stolen, purloined or converted . . . shall be fined under this title or imprisoned not more than ten years, or both.[45]

Isn't this precisely what Clinton did? She converted government documents for her use on a private server in her home. Only after she was out of office and under legal pressure from the Benghazi Committee did she give some of the documents to the State Department nearly two years later. That's like stealing a car, but only giving it back after being caught by authorities red-handed. The return of the car does not vitiate the crime.

But Clinton also concealed, removed, and destroyed government documents and wiped her private record-keeping server clean by using a file-deleting software even though she had been instructed by Congress to preserve all documents. The FBI later discovered that thousands of work-related documents had never been turned over by her to the federal government, thus depriving the government of its property. On this basis, Clinton violated 18 U.S.C. 2071(b):

> Whoever, having the custody of any such record, proceeding, map, book, document, paper, or other thing, willfully and unlawfully conceals, removes, mutilates, obliterates, falsifies, or destroys the same, shall be fined under this title or imprisoned not more than three years, or both.[46]

Clinton created a private server for all her official communications as secretary of state. In so doing, she purposefully deprived the government of the documents that belonged not to her, but to the government.[47] Case law supports a criminal prosecution under these facts.[48]

But there is more. By her own admission, Clinton directed that all her emails be deleted. Although Clinton had promised to cooperate fully with the Benghazi Committee and its demand to preserve and produce documents, her lawyer, David Kendall, subsequently sent a letter to the House Benghazi Committee on March 27, 2015, confirming that both personal and business emails had been removed from his client's server and backup systems.[49]

Making a materially false statement to Congress is a crime under 18 U.S.C. 1001.[50] It is also obstruction of justice under 18 U.S.C. 1505 and 1515(b) to act "corruptly" by "withholding, concealing, altering, or destroying a document or other information" in a congressional investigation.[51] Not all of Clinton's emails were produced as the committee had demanded and as Clinton had represented. The act of wiping her server clean of all original documentary evidence sought by congressional committee meets the legal requirements of these crimes.

Significantly, Clinton broke other laws involving the very serious matter of mishandling of classified documents. In his statement to the national media on July 5, 2016, FBI director Comey revealed his findings:

> From the group of 30,000 emails returned to the State Department, 110 emails in 52 email chains have been determined by the owning agency to contain classified information at the time they were sent or received. Eight of those chains contained information that was Top Secret at the time they were sent; 36 chains contained Secret information at the time; and eight contained Confidential information, which is the lowest level of classification. Separate from those, about 2,000 additional emails were "up-classified" to make them Confidential; the information in those had not been classified at the time the emails were sent.[52]

Comey's statement should have been the equivalent of an indictment of Clinton under 18 U.S.C. 793(f):

> Whoever, being entrusted with or having lawful possession or control of any document . . . relating to national defense, (1) through gross negligence permits the same to be removed from its proper place of custody . . . or (2) having knowledge that the same has been illegally removed from its proper place of custody . . . shall be fined under this title or imprisoned not more than ten years, or both.[53]

Clinton surely *knew* that if her emails were stored on her private server in the basement of her home, there would inevitably be innumerable classified documents residing in an unauthorized place. This very clearly jeopardized national defense as the statute envisions. At the very least, her conduct was grossly negligent as the above statute requires.

But beyond gross negligence, Clinton's handling of classified information appears to have been quite intentional. She *intended* to establish a nongovernment server and *intended* that it be used for all her government business, which necessarily included classified documents. Such willful and deliberate retention is defined as criminal under 18 U.S.C. 793(d) and (e).[54] This will be explored in greater detail in the next chapter.

A separate law also criminalizes Clinton's actions. 18 U.S.C. 1924(a) prohibits the following:

> Whoever, being an officer . . . of the United States, and, by virtue of his office, employment, position, or contact, becomes possessed of documents or materials containing classified information of the United States, knowingly removes such documents or materials without authority and with the intent to retain such documents or materials at an unauthorized location shall be fined under this titled or imprisoned for not more than five years, or both.[55]

The facts show that Clinton intended to create a private server. She knew it was unauthorized. Yet, she deliberately used it for all her electronic communications, including classified documents. These actions are violations of the above-stated statutes. Other people in government have been prosecuted and convicted with much less incriminating evidence, as will be explained in the next chapter.

In addition to the potential felony charges outlined, Clinton could also be charged with multiple conspiracy counts since her actions involved acting in concert with other participants as defined in 18 U.S.C. 371 and 18 U.S.C. 286.[56]

CLINTON'S PROTECTORS

The evidence shows that Clinton felt privileged and above the law. She wanted to control the records of her high office and keep them secret at any cost. She thwarted public access to nonclassified documents. In the process, she left vital classified information, including "top-secret" material, vulnerable to theft by hackers, foreign intelligence services, and governments seeking to steal America's national security information.

Clinton's principal defense of her actions was, in truth, no defense at all. She insisted that she utilized a private server "for convenience" as if that was somehow a legal excuse for violating multiple criminal statutes.[57] The law abides no such defense, as Lord Justice Pearce observed long ago. She also attempted to shift the blame or minimize her responsibility by claiming that others in her position had done the same thing.[58] That is a fatuous argument. It is a little like saying that other people have robbed banks, so it is okay for me to rob a bank.

While it is true that some government officials had in the past used private email accounts for some occasional official business, the practice was rare. None of Clinton's predecessors created a private server for *all* their communications as secretary of state. The reasons were quite simple. The law imposes strict rules limiting such use. First, the emails must be carefully preserved by the employee's agency.[59] Second, the communications cannot be destroyed by the employee.[60] Third, and most significantly, no classified material can be maintained on any private account.[61]

Clinton managed to violate these provisions of law during her four-year tenure as secretary of state. The public was unaware because there was no one in authority at the State Department to oversee whether Clinton was complying with the law. This would have been the duty of an internal inspector general who is charged with uncovering mismanagement and corruption. However, during the entire time of Clinton's service, President Barack Obama never nominated anyone to fill the position.[62] While there was an "Acting Inspector General," there was never a permanent I-G who might have exercised greater authority to uncover how Clinton had circumvented the law. This was likely a deliberate arrangement furtively negotiated by Clinton. Any other explanation was implausible.

Joseph diGenova, a former U.S. Attorney for the District of Columbia who also served as an independent counsel, was blunt in his assessment of how Clinton avoided detection:

> She was surrounded by a group of people at senior levels who protected her and protected the information. And, of course, it wasn't on government servers, so there wasn't an easy way for people in government to find out about it.
>
> So, she had a scheme, she had a methodology, and they had made sure that a watchdog was not present as long as she was Secretary of State. It was brilliantly carried out and it was accomplished with the complicity of the president who agreed not to appoint a permanent Inspector General.
>
> The president and the people in the White House knew she had a private server and they didn't object.[63]

Doug Burns, who served as assistant U.S. Attorney for the Eastern District of New York, concluded that Clinton broke the law and should have been indicted for several crimes:

> The establishment of the private server and its use with classified information plainly violated the law, all of her excuses notwithstanding. And to those excuses, the withholding, wiping and destruction of such huge amounts of email evidence removes all doubts about intent and knowledge.[64]

Given all the compelling evidence of criminality, how can it be explained that Hillary Clinton managed to escape prosecution? It appears she had considerable help.

CHAPTER 2

COMEY CONTORTS THE LAW TO CLEAR CLINTON

It is a public scandal when the law is forced to uphold a dishonest act.

—Lord Edward MacNaghten, House of Lords (1894)

After a year of gathering incriminating evidence of wrongdoing, FBI director James Comey held a news conference on July 5, 2016, to announce the findings of his investigation.

Across the country, many of those closely following the case expected him to outline evidence of a series of crimes committed by Hillary Clinton for the Department of Justice to pursue. After all, many top officials who committed similar transgressions on a smaller scale saw criminal charges that ended their political careers.

It was one of the most bizarre, bewildering, and incomprehensible explanations of a legal case in modern American history. Comey detailed evidence of crimes but did not refer her for prosecution. Comey's assessment of the law was not just flawed, but completely inaccurate and conspicuously wrong. His reasoning for not recommending prosecution of Clinton was beyond improbable. It was impossible.

Just as Lord MacNaghten feared, Comey bent and distorted the law to uphold dishonest acts by Clinton.

COMEY'S TORTURED INTERPRETATION

In front of a national media, Comey read a lengthy statement that presented a compelling case of how Clinton repeatedly violated the law: 110 classified emails were stored by her on an unsecured private server in her home without authorization and "none of these emails should have been on any kind of unclassified system."[1]

Comey dismissed as untrue Clinton's ever-changing excuses that classified documents were marked in a way that she would not recognize them as such. "But even if information is not marked 'classified' in an email, participants who know or should know that the subject matter is classified are still obligated to protect it," he observed. For example, Comey cited seven email chains involving top-secret communications, stating, "There is evidence to support a conclusion that any reasonable person in Secretary Clinton's position . . . should have known that an unclassified system was no place for that conversation."[2]

Clinton's routine failure to protect America's classified information jeopardized national security. "We assess it is possible that hostile actors gained access to Secretary Clinton's personal email account," Comey concluded.

Having laid out a meticulous case of how Clinton broke the law, Comey suddenly offered an implausible reason for why she should not be prosecuted:

> Although there is evidence of potential violations of the statutes regarding the handling of classified information, our judgment is that no reasonable prosecutor would bring such a case.[3]

The first part of Comey's finding that "there is evidence of potential violations of the statutes" was more than legally sufficient to refer the case to the Justice Department for criminal prosecution and presentment of evidence before a grand jury to seek an indictment. He unambiguously asserted facts which would meet any standard of probable cause that crimes appeared to have been committed by Clinton. On this basis alone, Comey was duty-bound to tender a criminal referral for prosecution. No subsequent qualifying predicate or condition to the statement can eliminate that legal requirement under the law.

The second part of Comey's statement that "no reasonable prosecutor would bring such a case" was nothing more than speculation by him personally. The law does not accept this as a valid basis for declining to levy charges. Moreover, Comey had no grounds for making such a bold claim. He appeared to have simply invented it. In truth, plenty of prosecutors would have been willing to "bring such a case" and, indeed, many have done so under similar circumstances.

All the former FBI officials interviewed for this book said they had never heard of a decision being based on what "a reasonable prosecutor" might or might not do. Danny Coulson, who served as deputy assistant director of the FBI during his three decades at the bureau and is also a lawyer, stated:

> My view, and the view of agents in the field, was shock. When Comey came to the end and said no reasonable pros-

> ecutor in the entire country would prosecute Clinton, that was a total lie. I was flabbergasted. The agents were, too. It didn't make any sense to any of us.
>
> There was never an FBI investigation of Hillary Clinton. Comey controlled it from start to finish and came out with the results he wanted.[4]

Former assistant FBI director Steve Pomerantz was equally shocked and appalled by what he witnessed the day Comey made his televised announcement:

> I could have fallen off my chair while watching that news conference. Setting aside the conclusion he drew, it is not the FBI's job to recommend prosecutions. The FBI investigates and then turns it over to the Department of Justice. In all my years in the FBI—over thirty years—and hundreds of investigations, probably thousands, I never ever saw that done. For Comey to do that was astonishing and wrong.[5]

Comey's assertion that "no reasonable prosecutor would bring such a case" was remarkably inaccurate. In truth, federal prosecutors have brought numerous cases against individuals who mishandled classified documents in much the same way that Clinton did.

Samuel "Sandy" Berger, who served as national security adviser under President Bill Clinton, removed classified documents from the National Archives and stored them in an unauthorized location. Like Clinton, Berger said he did it for convenience.[6]

David Petraeus, who was director of the Central Intelligence Agency under President Barack Obama, kept classified documents at his home and allowed his biographer, with whom he was having an affair, to read them.[7] Both Berger and Petraeus faced criminal charges and pleaded guilty. Both were fined, put on several years' probation, and stripped of their security clearances.

John Deutch served as CIA director under President Clinton. Classified materials, including top-secret documents, were found on a computer in his home. While negotiating a plea agreement with prosecutors, he received a presidential pardon.[8]

Crucially, Petraeus, Berger, and Deutch never worked for the government again. Even though none of them served time behind bars, their prosecutions and inability to maintain security clearances meant that the holding of high office was an impracticality. Had Clinton faced the same legal standard for mishandling classified documents, her ambition to serve as president would have ended.

Other individuals who did not hold public office have also been prosecuted for the same actions as Clinton. Navy engineer Bryan Nishimura downloaded classified documents on an unsecured computer system and stored them on personal electronic devices.[9] He also destroyed some of the evidence. Like Clinton, he insisted that he had no "intent" to break the law. Nevertheless, he was charged and pleaded guilty in 2015.

In 2016, a young navy sailor, Petty Officer First Class Kristian Saucier, was charged with mishandling classified material when he took six photographs of the inside of a submarine with his personal cell phone, an unauthorized and unsecured device.[10] He was eventually forced to plead guilty, even though his attorney argued what he called "the Clinton defense," claiming that if Clinton could get away with far worse conduct, his client should not be found guilty.[11]

The fact that the "Clinton defense" did not work for Saucier suggests that it would not have succeeded for Clinton. But, of course, she was never even charged, unlike the sailor and several other troops. In March 2018, having served one year in prison,

Saucier received a presidential pardon. Trump had previously noted that the sailor had been held to an unfair "double standard" compared to Clinton, who managed to escape prosecution entirely.[12]

During President Obama's term, more than half a dozen people were charged and convicted of mishandling classified materials.[13] Nearly all of them served time behind bars. Thus, Comey's assertion that "no reasonable prosecutor would bring such a case" was demonstrably untrue.

Proof of these prior cases of a similar nature made another of Comey's legal pronouncements even more bewildering. In his July 5 statement, he advanced this specious claim:

> In looking back at our investigations into mishandling or removal of classified information, we cannot find a case that would support bringing criminal charges on these facts. All the cases prosecuted involved some combination of: clearly intentional and willful mishandling of classified information; or vast quantities of materials exposed in such a way as to support an inference of intentional misconduct. . . .[14]

The prosecutions of Berger, Petraeus, Deutch, Nishimura, and Saucier disproved the first part of Comey's statement since they all bear a notable resemblance to Clinton's case. Moreover, at the time Comey cleared Clinton, the FBI and the DOJ were already in the process of building a criminal prosecution against Harold T. Martin III, a former national security contractor who stored classified documents in his home—some of which were on his computers—just as Clinton did. Martin was arrested, charged, and jailed one month after Comey absolved Clinton for doing the same thing.[15]

The number of classified materials found at Martin's home was described by federal prosecutors as "astonishing." The same could be said of Clinton's tranche of prohibited records. Her computer server held not just a handful of classified documents, but "110 emails in 52 email chains . . . determined by the owning agency to contain classified information at the time they were sent or received. . . . Separate from those, about two thousand additional emails were 'up-classified,'" meaning they were later determined to contain classified information that someone in Clinton's position should have recognized immediately upon reading them.[16]

Former deputy assistant director of the FBI Danny Coulson believes that Comey's statement that he knew of no other comparable cases that would support charging Clinton shows the director was either thoroughly uninformed or severely biased in favor of the ex-secretary of state:

> All of these individuals were prosecuted for doing a lot less than Clinton did. I don't know how Comey can look at those cases and look at what he recommended for her and not see the inconsistency, unless you come to the conclusion that he got the result he preconceived.[17]

THE CRIMINAL "INTENT" CANARD

In his quest to exonerate Clinton of any wrongdoing, Comey decided to focus on the suspect's "intent" and then proceeded to misconstrue the law that governs intent.

His tortured logic was set forth in this key sentence during his July 5 announcement:

> Although we did not find clear evidence that Secretary Clinton or her colleagues intended to violate laws governing the

> handling of classified information, there is evidence that they were extremely careless in their handling of very sensitive, highly classified information.[18]

First, nowhere in 18 U.S.C. 793(f) of the Espionage Act which governs the "grossly negligent" handling of classified information does it state that a defendant must have intended to break the law in order to be charged or found guilty. In clearing Clinton on this basis, Comey inferred language that does not exist. He either misinterpreted the law or chose to rewrite it.

The plain language of the statute is unmistakable: no intent is required. It states, "Whoever, being entrusted with or having lawful possession or control of any document . . . relating to national defense, (1) through gross negligence permits the same to be removed from its proper place of custody" has committed a felony.[19] Unlike other sections of the same statute, the word "willingly" was omitted. This was a deliberate decision by Congress when it revised and expanded the Espionage Act decades after it was enacted. Section (f) was added to provide a lesser alternative to willful conduct; that is, grossly negligent behavior.[20]

Second, the other sections of 18 U.S.C. 793—except provision (f)—do, indeed, have an "intent" or "willful" requirement. They state that willfully or intentionally retaining classified information in an unauthorized place or transmitting it to an unauthorized person is a felony.[21] Clinton also violated these provisions of law in two ways: by intentionally storing classified emails in an unauthorized place (her home) knowing they would be vulnerable to foreign access and by intentionally giving all of her emails, including classified documents, to her lawyers and others for sorting. They did not appear to have received classified security clearance to review them.

Despite all of this, Comey once again misapplied the law of "intent" in clearing Clinton. The plain meaning of the statutory language and the accepted legal standard is not whether Clinton intended to violate the law, as Comey claimed, but whether her *actions* were intentional. Unquestionably, they were.

Even if one were to accept Comey's interpretation, the statute also makes it a crime if the possessor of classified documents has "*reason to believe* (it) could be used to the injury of the United States or to the advantage of any foreign nation" if it is kept in an unsecured place.[22] How was it possible that Clinton did not have such a reasonable belief? She must have. Anyone in government knows all too well how classified documents must be protected from foreign theft. Failure to safeguard them could cause injury to the U.S. Under the Espionage Act, it is not necessary to prove Clinton intended to jeopardize national security or to aid a foreign power; only that she had reason to believe it could occur as a consequence of her deliberate actions.[23]

This is why Clinton was instructed on her first day on the job that all such materials must be maintained in a government-approved and secure location. Her home was not. Clinton knew this and, therefore, had "reason to believe" her actions would jeopardize national security. Remember, Comey openly admitted this in his July 5 statement when he said, "there is evidence to support a conclusion that any reasonable person in Secretary Clinton's position . . . should have known that an unclassified system was no place for that conversation," referring to seven email chains involving Top-secret communications.[24]

Clinton didn't just *keep* classified documents in an unauthorized place, but she willingly *gave* them to people she knew did not have security clearance to receive them, in violation of the law. Comey acknowledged that his investigation had discovered this when he was cross-examined by Representative Jason Chaffetz during a hearing before the House Oversight Committee a mere two days after he cleared Clinton:

> **CHAFFETZ**: Did Hillary Clinton give noncleared people access to classified information?
> **COMEY**: Yes. Yes.[25]

According to Representative Trey Gowdy during the hearing, Clinton gave "up to ten people" access to classified information, and none of them were permitted to have such protected materials.[26] It is truly baffling why she was not prosecuted on this basis alone.

Third, the Espionage Act is not the only law that criminalizes the mishandling of classified documents. 18 U.S.C. 1924(a) makes it a crime to knowingly remove classified records without authority "and with the *intent* to retain such documents or materials at an unauthorized location."[27] The language of this law properly states the legally accepted "intent" standard. It does not require that Clinton intend to violate the law, as Comey stated. Since she never sought authorization for the server that stored them, she should have been charged under this statute.

During her March 10, 2015, news conference, Clinton admitted that she intentionally used her personal server for the sake of convenience.[28] She deliberately refused to utilize a government email account on a secure server hosted by the Department of State, and her top staff fought efforts to have her do so, as the FBI confirmed.[29] These are the intentional acts that placed 110 classified documents in a vulnerable location and that, in the end, exposed America's national security secrets to foreign intrusion.

As explained in chapter 1, Clinton had been told at the outset of her service as secretary of state that this was against the law, and she acknowledged she understood it when she signed the document entitled "Classified Information Nondisclosure Agreement."[30] The evidence is conclusive that Clinton knew she was committing crimes.

CLINTON WAS GROSSLY NEGLIGENT

Setting up an unclassified email server that houses classified material is the definition of grossly negligent behavior under the Espionage Act. And the sheer *quantity* of classified information—110 documents—is evidence of gross negligence. Yet Comey managed to either ignore the law or turn its meaning upside down. He described Clinton's actions as "extremely careless." What is the difference? There is none.

In courtrooms across America, the terms "gross negligence" and "extremely careless" are devoid of any difference and, hence, synonymous. When juries struggle with the meaning of "gross negligence," judges will often explain that it is behavior of a reckless or extremely careless nature. By describing Clinton's conduct as "extremely careless," Comey was admitting that Clinton broke the law, although he twisted the language and the statute to create the appearance that she did not.

Former federal prosecutor Andrew McCarthy explained it this way: "Substantively, these terms are indistinguishable. The emendation is said to be critical, though, because the statute applicable to Clinton's conduct criminalizes 'gross negligence.' If Comey had said the words 'grossly negligent,' so the story goes, it would be the equivalent of pronouncing Clinton guilty. That is, it would be impossible to rationalize not charging her because, after all, this was all about negligence."[31] Instead, Comey cleverly altered the language to create the public *impression* or *illusion* that Clinton had barely skated across the thin ice of illegality when, in truth, she had plunged right through.

Comey must have known Clinton committed felonies because an early version of his so-called exoneration statement written in May 2016 did, in fact, use the words "grossly negligent" to describe Clinton's handling of classified information. Here is how he recounted Clinton's conduct in the original draft:

> There is evidence to support a conclusion that Secretary Clinton, and others, used the email server in a manner that was grossly negligent with respect to the handling of classified information.[32]

This finding would have led to a criminal referral for prosecution of Clinton and an inevitable indictment. But it never happened. And almost no one knew this is what Comey had first determined.

Two years after Comey scuttled the case against Clinton, documents uncovered by John Solomon of *The Hill* showed that the words "grossly negligent" had been edited out with red lines on or around June 6, 2016, three weeks before the FBI interviewed Clinton.[33] The term "extremely careless" was substituted. These alternative words, on their face, seemed less culpable and more innocuous. But the law views them as identical to "gross negligence." Thus, it appears that Comey sought to draw a fine distinction where none exists. By avoiding the exact language of 18 U.S.C. 793(f), Comey managed to exculpate Clinton by employing a legal charade.

Documents made available by the Senate Homeland Security Committee also show that Comey originally found another basis to conclude that Clinton broke the law. In his draft statement, he noted the following:

> Similarly, the sheer volume of information that was properly classified as Secret at the time it was discussed on email (that is, excluding the "up-classified" emails) supports an inference that the participants were grossly negligent in their handling of that information.[34]

This finding reinforced the conclusion that Clinton committed crimes. In other words, there were so many classified documents on her unclassified system that she had to have been "grossly negligent" under the criminal statute. However, this part of Comey's statement was also later modified in a critical way. The entire reference to the "inference" of criminality was deleted.

Former federal prosecutor Doug Burns believes Comey fully realized that Clinton committed several felonies, but decided to rewrite the law and reshape the facts to clear her:

> It was beyond embarrassing to see the director of the FBI twist himself into a pretzel trying to explain the non-prosecution decision in the Clinton matter. He violated every rule in the book. Agents do not make prosecutorial decisions and prosecutors would never lay out the proof of the crime and then bizarrely explain that no case is being brought. His explanations about intent and carelessness were wrong legally and, again, totally embarrassing.[35]

DID COMEY OBSTRUCT JUSTICE?

These actions by Comey raise the specter that the FBI director may have been motivated to absolve Clinton for reasons that have little to do with the law and the facts. If so, this could constitute obstruction of justice.

The best evidence that shows Comey cleared Clinton for political reasons is that he wrote his statement exonerating her several weeks before all the facts were known. According to the Senate Judiciary Committee, transcripts of interviews with people close to the FBI director prove that Comey drafted his statement well in advance of the FBI's interviews with Clinton and sixteen other witnesses.[36] How can that be? Comey could not possibly have known what legal conclusions to draw in the absence of fundamental facts that only those key participants could provide, especially the subject of his investigation, Clinton. Comey must have been determined to absolve Clinton regardless of what she or others had to say under FBI questioning.

Fred Tecce, who was an assistant U.S. attorney in Philadelphia and, later, special assistant U.S. attorney at the Department of Justice, believes Comey threw the case to protect Clinton:

> The Investigation was a sham, a farce. They wrote the press release announcing the results of the investigation before the investigation fully took place. The fact that they even bothered to interview her was a joke. Clinton could have told them "yeah, I stole everything, the computer was illegal, I knew what I was doing was wrong" and they would have said, "thank you very much madam secretary, have a nice day," and then they would have issued the press release they had written weeks before announcing that there wasn't enough evidence to prosecute. It was a joke.[37]

All the former federal prosecutors and former top officials at the FBI who were interviewed for this book said they had never seen such a practice. They described it as wrong and in violation of FBI standards. The Senate Judiciary Committee accused Comey of predetermining the outcome of the case against Clinton:

> . . . Mr. Comey had already decided he would issue a statement exonerating Secretary Clinton. That was long before FBI agents finished their work. Mr. Comey even circulated an early draft statement to select members of senior FBI leadership. The outcome of an investigation should not be prejudged while FBI agents are still hard at work trying to gather facts.
>
> Conclusion first, fact-gathering second—that's no way to run an investigation. The FBI should be held to a higher standard than that, especially in a matter of such great public interest and controversy.[38]

Why would Comey change the crucial language to clear Clinton in the face of incriminating evidence that she mishandled classified material by both intentional acts and through grossly negligent behavior? Did Comey absolve Clinton for political reasons? Was he pressured to do so by Attorney General Loretta Lynch or others at the Justice Department? Did top officials at the FBI, motivated by corrupt purposes, dissuade him from making the criminal referral which his *original* draft statement so plainly set forth? And why exactly did he take it upon himself to usurp the authority of the Justice Department in clearing Clinton?

If Comey prevented the prosecution of Clinton in the face of sufficient and compelling evidence to indict her or was complicit in doing so, as the evidence indicates, then such interference or intervention in the due administration of justice would rise to the level of obstruction of justice by Comey.

18 U.S.C. 1505 makes it a crime to obstruct justice:

> Whoever corruptly . . . influences, obstructs, or impedes or endeavors to influence, obstruct, or impede the due and proper administration of the law under which any pending proceeding is being had before any department or agency of the United States . . . shall be fined under this title, imprisoned not more than 5 years, or both.[39]

The operative word in the statute is "corruptly." What does it mean? Courts have struggled with its meaning over the years, which is why Congress decided to define it in 18 U.S.C. 1515(b) as "acting with an improper purpose, personally or by influencing another, including making a false or misleading statement, or withholding, concealing, altering, or destroying a document or other information."[40]

Thus, clearing Clinton for a personal or political reason could easily amount to a "corrupt" or improper purpose constituting obstruction of justice. So, too, would the act of altering an exoneration statement with a false finding of "extremely careless" if the facts determined that Clinton's actions were "grossly negligent."

Comey was asked, under oath, by the House Judiciary Committee if he decided not to pursue criminal charges against Clinton *before* or *after* she was interviewed by the FBI. He testified "after."[41] However, the documents uncovered by the Senate Judiciary Committee belie his testimony. A full two months before the FBI ever interviewed Clinton and her top aides, Comey began drafting the statement that he used to exonerate Clinton. Absent some extraordinary explanation, it appears that Comey's investigation was little more than a charade and that he may have lied under oath. If true, it would constitute the crime of perjury under 18 U.S.C. 1621 or a false statement under 18 U.S.C. 1001.[42]

This document establishes persuasive evidence that Comey

preordained that Clinton would not be charged. What prosecutor writes a statement absolving a suspect before the evidence is fully gathered, especially from the principal witnesses? No prosecutor of which I am aware. Unless, of course, the fix was in. Unless someone instructed or encouraged him to protect Clinton or he decided to do it all on his own with a presidential election hanging in the balance.

Either way, it is evidence of obstruction of justice. It is a felony to interfere with a criminal investigation, even if you are in charge of it. It is also illegal to use your public office for a political purpose, if that is what Comey did.

Former FBI assistant director Bill Gavin, who spent twenty-eight years with the bureau, said he had never heard of anyone in the FBI, including the director, composing an exoneration statement prior to the conclusion of an investigation. He referred to it as "unprecedented and wrong."[43] Former assistant deputy director Buck Revell agreed, calling it inconsistent with the duties and responsibilities of the FBI:

> That is totally outside the scope of any reasonable course of action. It never happened to my knowledge, and I ran operations for ten years. I know what happened back then. It would have been something that caused me to go to the Congress and say that we've got a conflict of interest in our own organization.[44]

The questions then become two-fold: Did Comey do this on his own? Did others exert pressure or unduly influence him to exculpate Clinton? Facts later emerged suggesting that the investigation was corrupted by several people from within the FBI and the Justice Department.

LORETTA LYNCH AND A SERIES OF FANTASTIC COINCIDENCES

We are expected to believe it was a coincidence that former President Bill Clinton just happened to be on the tarmac of Sky Harbor International Airport in Phoenix, Arizona, at exactly the same time as Attorney General Loretta Lynch on June 27, 2016, a scant five days before Hillary Clinton was to meet with FBI officials for questioning about her suspected wrongdoing. Perhaps it was also just a coincidence that eight days after the furtive tarmac meeting the decision was announced that criminal charges against Clinton would not be filed.

We are supposed to accept that the private meeting on board Lynch's plane had nothing whatsoever to do with the criminal investigation of Clinton that the AG was overseeing at the time and that it was a "primarily social" interaction. The optics, if not the substance, were beyond improper by any legal and ethical standard: the husband of the subject of a criminal investigation was meeting with the one person who could decide whether charges would be brought against his wife.

Had a local reporter not been tipped off about the rendezvous, no one would have ever known about their discussion that reportedly lasted for half an hour.[45] The secret meeting would have remained a secret and, in all likelihood, Lynch would not have felt compelled to accept whatever recommendation was made by Comey.

Lynch should never have presided over the Clinton email case to begin with. She owed her career to none other than Bill Clinton, who nominated her to serve as U.S. Attorney for the Eastern District of New York, which nicely positioned her for elevation to attorney general a few years later. She should have recused herself from the Clinton probe at the outset, but never did so. Even when the ethically challenged tarmac meeting took place, she refused to recuse herself, saying she would accept the recommendations of the FBI.[46]

Lynch was *required* under law to disqualify herself. Under the Code of Federal Regulations, 28 CFR 45.2, "no employee shall participate in a criminal investigation or prosecution he (or she) has a personal or political relationship with any person or organization substantially involved in the conduct that is the subject of the investigation or prosecution."[47] Lynch had *both* a personal and professional relationship with Bill and Hillary Clinton. The private tarmac meeting underscored their close personal connection, and Bill's selection of Lynch for a high-ranking position at the Justice Department is evidence of their professional association.

The Democratic-led Justice Department's objectivity and independence were already being questioned when the meeting occurred. But we are supposed to believe the Lynch-Clinton encounter was just an innocent coincidence of a casual, social nature. Why, then, did the FBI on the scene instruct everyone that "no photos, no pictures, no cell phones" could be used to capture what took place?[48]

When the confab became public, the American Center for Law and Justice (ACLJ) requested any and all records of the meeting and were told by the FBI that "No records responsive to your request were located."[49] A year later, the DOJ produced more than four hundred pages of emails, although the contents of what Lynch and Clinton discussed were heavily redacted. The emails do reveal that reporters for the *New York Times* and the *Washington Post* seemed inclined to minimize the matter as an innocuous nonstory and reached out to Justice Department officials to reinforce that view.[50]

Comey, for his part, told Congress almost a year later that he became suspicious of Lynch's motives and lost faith in the Justice Department's investigation following the tarmac encounter. This, he said, became his justification for deciding, on his own, to assume prosecutorial authority over the probe, although the evidence shows he had already taken steps to assume command even before the meeting.

Testifying before the Senate Judiciary Committee on May 3, 2017, Comey made a cryptic statement about impropriety at the Department of Justice:

> A number of things had gone on which I can't talk about yet, that made me worry that the department leadership could not credibly complete the investigation and decline prosecution without grievous damage to the American people's confidence in the justice system.[51]

Comey was asked in the hearing about an email that reportedly provided assurances that Lynch would protect Clinton by making sure the FBI investigation "didn't go too far." The director responded, "That's not a question I can answer in this forum . . . because it would call for a classified response. I have briefed leadership of the intelligence committees on that particular issue, but I can't talk about it here."[52] It is doubtful that his response would involve anything classified. It appeared to be an excuse to keep the matter hidden, as it remains.

A month later, Comey raised even more suspicions about Lynch and potential impropriety when he told the Senate Intelligence Committee that the then-attorney general had instructed him to mislead the public about the true nature of the Clinton investigation. Comey testified, "The Attorney General had directed me not to call it an investigation, but instead to call it a matter, which confused me and concerned me."[53] In follow-up questions, Comey said Lynch's request "concerned me because that language tracked with how the (Clinton) campaign was talking about how the FBI was doing its work."[54] In Comey's mind, Lynch was not evaluating the law and facts objectively, as was her duty. Instead, she was actively attempting to help Clinton beat the rap.

Comey's self-serving explanations for recommending that Clinton not be prosecuted were contrary to the law and raised more questions than answers. If he believed that Lynch was biased and determined to protect Clinton, why didn't he take more aggressive action to expose it? He should have immediately notified leaders of Congress or the Justice Department's inspector general. He did neither, choosing to remain silent. If he felt that there was corruption born of a conflict of interest and nothing was being done to correct it, he should have resigned his post and exposed his suspicions to the public. Comey's failure to act suggests he had his reasons for commandeering complete authority over the Clinton case in order to accomplish the outcome he had already predetermined in her favor.

Former U.S. Attorney Joe diGenova, who also served as an independent counsel, surmised that Comey and Lynch were motivated by a desire to see Clinton become president:

> What Comey and Lynch did was to corrupt the statute to save Clinton. Everything that Comey said at the news conference was false. Clinton had clearly violated the Espionage Act. There just isn't any doubt about that. For him to say that no reasonable prosecutor would bring the case was ludicrous because reasonable prosecutors have been bringing cases over far less than that for years. Comey knew that. He

became an embarrassment that day because everything he said was false. And he did it for one reason, to help Clinton become president.[55]

COMEY'S DECISION ANGERED FBI AGENTS

FBI director Comey's decision to let Hillary Clinton off the hook proved to be an extremely unpopular one within the bureau. A senior FBI official close to the investigation told Fox News that "career agents and attorneys on the case unanimously believed the Democratic presidential nominee should have been charged" and the decision "left members of the investigative team dismayed and disgusted."[56] The October 13, 2016, story revealed that more than one hundred FBI agents and analysts developed the evidence, along with six attorneys from the Justice Department's national security division, counterespionage section, who reviewed what laws Clinton broke.

"No trial level attorney agreed, no agent working the case agreed, with the decision not to prosecute—it was a top-down decision" to decline to bring charges against Clinton, the insider confided. In other words, Comey and several of his top aides, perhaps in collaboration with the DOJ, chose not to pursue an indictment. "It is safe to say the vast majority felt she should be prosecuted," said the source. "We were floored while listening to the FBI briefing because Comey laid it all out, and then said, 'but we are doing nothing,' which made no sense to us."[57]

It was especially confounding because a grand jury was never fully utilized, nor was there any indication that search warrants were issued. All of this would be customary in a case involving potential national security breaches caused by the mishandling of classified information. These departures from normal procedure were obvious red flags and deeply alarming to the agents who were working the case. Thus, the entire process seemed distorted and illegitimate from the outset, as if the investigation was nothing more than a subterfuge designed to cover up illegality.

A story from Fox News also pointed out a strange and suspicious twist:

> Another oddity was the five so-called immunity agreements granted to Clinton's State Department aides and IT experts. Cheryl Mills, Clinton's former chief of staff, along with two other State Department staffers, John Bentel and Heather Samuelson, were afforded immunity agreements, as was Bryan Pagliano, Clinton's former IT aide, and Paul Combetta, an employee of Platte River networks, the firm hired to manage her server after she left the State Department.
>
> As Fox News has reported, Combetta utilized the computer program Bleachbit to destroy Clinton's records, despite an order from Congress to preserve them, and Samuelson also destroyed Clinton's emails. Pagliano established the system that illegally transferred classified and top-secret information to Clinton's private server. Mills disclosed classified information to the Clinton's family foundation in the process, breaking federal laws.[58]

Why were these five people given immunity from prosecution? In almost every criminal case, immunity is only granted after a witness delivers a "proffer" (an offer of proof or evidence) that incriminates someone else and precipitates criminal charges. Yet no one, Clinton included, was ever prosecuted.

The prospect of a Comey cover-up was further fueled by the inexplicable actions of the FBI when it reportedly destroyed the laptops of Samuelson and Mills after they received immunity. Why would the FBI erase or demolish computers with classified information contained therein? It appears the Bureau itself committed crimes by destroying evidence relevant to its own criminal investigation.

But there's more. According to the senior FBI source, "Mills was allowed to sit in on the interview of Clinton as her lawyer. That's absurd. Someone who is supposedly cooperating against the target of an investigation (being) permitted to sit by the target as counsel violates any semblance of ethical responsibility."[59]

Indeed, it does.

Samuelson, a lawyer, was also permitted to be present when Clinton testified. It was not only irregular, but highly improper, unethical, and probably illegal. Under 18 U.S.C. 207, it is against the law for former government officials, like Mills and Samuelson, to "knowingly make, with the intent to influence, any communication before any officer or employee of any department or agency of the U.S. in connection with a particular matter" they were once personally involved as government employees.[60]

It seems that there was never any serious possibility that Clinton would be charged for anything she might admit to in her July 2, 2016 interview. Otherwise, the FBI would not have allowed two immunized witnesses central to the prosecution's case to sit in the same room to hear her testimony. It is unheard of for prosecutors to permit such a thing because it would give those witnesses a chance to conform their testimony in such a way as to support the subject of the investigation, Clinton. During questioning before the House Judiciary Committee, Comey appeared to concede this point.

It was bad enough that the immunity deals called for the FBI to destroy the laptops of Mills and Samuelson, thereby obliterating valuable proof that Clinton may have destroyed evidence in purging her server. But it was far worse for the FBI to treat Mills and Samuelson as suspects, accomplices, witnesses, and defense lawyers all at the same time and in the same case. By condoning unethical behavior, the bureau simultaneously compromised the integrity and effectiveness of its own criminal probe. Of course, the FBI knew that. Somehow, they didn't seem to care. There was no stopping "the fix" once it was well under way.

FBI agents were also angry that Clinton was interviewed for only three and a half hours on July 2.[61] She reportedly answered, "I do not recall" or "I don't remember" some thirty-nine times.[62] At no point was she placed under oath. Comey cleared her of any wrongdoing three days later. "Every agent and attorney I have spoken to is embarrassed and has lost respect for James Comey and Loretta Lynch," the senior source told Fox News. "The bar for DOJ is whether the evidence supports a case for charges—it did here. It should have been taken to a grand jury."[63]

Several former FBI agents ventured damning assessments of both Clinton and Comey to Paul Sperry of the *New York Post*. Dennis V. Hughes, the first chief of the FBI's computer investigations unit, stated, "The FBI has politicized itself, and its reputation will suffer for a long time. I hold Director Comey responsible."[64]

Former FBI official I. C. Smith, who served as section chief in the National Security Division, observed, "Clearly there was a different standard applied to Clinton. What she did is absolutely abhorrent for anyone who has access to classified information."[65]

Retired agent Michael M. Biasello said, "Comey has single-handedly ruined the reputation of the FBI." Biasello claimed the outcome was predetermined "by design" and called Comey's decision "cowardly."[66]

When a decision makes no sense, it is reasonable to search for an alternative explanation or underlying reason that frequently exposes the sequestered truth.

That explanation came into focus when the inspector general of the Department of Justice, Michael Horowitz, decided to investigate Comey's mystifying decision. In the process, the IG uncovered the actions of a key FBI investigator by the name of Peter Strzok. And the name of FBI lawyer Lisa Page also emerged. Text messages exchanged between them led to the identities of others, including FBI deputy director Andrew McCabe, who appeared to have had personal and political motives to clear Clinton.

CHAPTER 3

"THE FIX"

Unlimited power is apt to corrupt the minds of those who possess it. Where law ends, there tyranny begins.

—William Pitt, prime minister of Great Britain, January 9, 1770

Very little was known about *how* FBI director James Comey reached his decision not to pursue criminal charges against Hillary Clinton.

Other than his public statement absolving her, nothing was revealed about the actions taken by agents and lawyers within the FBI who were intimately involved in the investigation and helped shape its improbable outcome. The Clinton exoneration statement and its evolution from criminality to absolution was a closely guarded secret within the agency. Americans were kept in the dark. So, too, was Congress.

Why did Comey initially find that Clinton was "grossly negligent" in violation of the law, but then alter the critical language to clear her of wrongdoing? How could that have even happened before any of the principal witnesses, including Clinton, were interviewed? It was puzzling, suspicious, and counterintuitive to the law. Had one or more individuals at the FBI put their thumb on the scales of justice to influence the result?

As troubling questions persisted over the ensuing months, several names began to emerge—Andrew McCabe, Peter Strzok, and Lisa Page. All three served as top FBI officials and were instrumental in the handling of the Clinton case. Together, they wielded enormous power—including the power to corrupt the process.

ANDREW MCCABE

During the early stages of the Clinton email investigation, McCabe was the assistant director in charge at the FBI's Washington field office. He was likely privy to the evidence being developed against Clinton. But in January 2016, nearly six months before Comey cleared Clinton of wrongdoing, he appointed McCabe to serve as deputy director of the FBI.[1] McCabe was second in command at the bureau and a close adviser to Comey. As such, he actively assumed oversight of the Clinton case and was intimately involved in the ultimate decision not to bring criminal charges against her. What few people knew at the time was how immersed McCabe had become in the same milieu as the Clintons.

In early 2015, Dr. Jill McCabe, a pediatric physician, was recruited by Democrats to run for the Virginia state senate just five days after Clinton's private email scandal became public.[2] Was this sequence of events a mere coincidence? Dr. McCabe's campaign coffers were the beneficiary of some $675,000 from groups aligned with Clinton and Virginia governor Terry McAuliffe, a longtime friend of Bill and Hillary Clinton and a former board member of the Clinton Global Initiative.[3] The timing of the payments is curious. The big dollar donations to McCabe arrived two months after the FBI launched its investigation of Clinton in July 2015.

McAuliffe personally solicited Dr. McCabe to run for office during a meeting that her husband also attended.[4] The subsequent cash donations were an unusually extravagant expenditure for a state-wide race and amounted to nearly 40 percent of her total campaign funds.[5]

The public knew none of this until three months after Clinton was cleared by Comey and his chief deputy, McCabe. The story broke on October 24, 2016, when Devlin Barrett reported it in the *Wall Street Journal.*[6] No one was ever able to prove that Clinton's fingerprints were on the bags of money, but McAuliffe has never denied that he helped arrange the cash.

Although there was severe criticism of McCabe within the FBI by fellow agents, the bureau officially insisted that he did not begin his management of the email probe until after his wife's campaign ended, and he only became involved in the case months later when he was promoted to deputy FBI director in January 2016.[7] Importantly, documents show that the FBI, knowing of his wife's cash connections to Clinton, never warned McCabe that he should stay clear of her email case, even after his wife lost her election.[8] Instead, he assumed an executive leadership role guiding the Clinton investigation in the critical six months preceding the FBI's announcement in July 2016 that she would not be criminally charged.

The FBI's defense of McCabe was both superficial and misleading. It makes no difference that his wife's campaign had ended. She still got the money. She was still indebted and beholden to Clinton and her close friend McAuliffe for their financial support of her. And Dr. McCabe's husband can hardly be described as an indifferent bystander with no interest in the matter. Spouses tend to support one another. That is why ethics advisers at the FBI cautioned him to recuse himself from all public corruption cases during his wife's senate race.[9]

The conflict of interest was glaring. But that conflict did not suddenly end at the conclusion of his wife's campaign. Dr. McCabe took money from political action groups closely associated with the person her husband ended up investigating. Common sense dictates that the McCabes' gratitude and the corresponding propensity for bias would persist in perpetuity. The decision to clear Clinton may have been influenced by that very bias.

This was a recurring problem throughout the investigation. Everyone involved was from a small circle of professional Washington insiders, who all knew each other well and have for years. It did not matter whether there was a quid pro quo. Surely, it would have been possible to find someone to determine Hillary Clinton's legal status in a fair and neutral manner.

At the very least, the appearance of impropriety should have been more than enough for McCabe to disassociate himself permanently from the entire criminal investigation of Clinton. Moreover, FBI director James Comey should have demanded it. He did not. The fact that Comey and McCabe declined to do so adds even more suspicion to the theory that "the fix was in" not to prosecute Clinton.

Documents show that McCabe did eventually recuse himself from the Clinton probe just days after the *Wall Street Journal* disclosed the flagrant conflict of interest—the same conflict of interest that may have allowed Clinton to escape prosecution. Only then, when the FBI was under siege by the public disclosure, did McCabe reluctantly disqualify himself, reportedly at Comey's urging, a mere one week before the presidential election on November 1,

2016.[10] At this point, of course, it was too late. His belated recusal was patently absurd since McCabe and Comey had already cleared Clinton and ended their investigation of her four months earlier.

McCabe's recusal came at a point when Comey briefly reopened the Clinton investigation on October 28, 2016, when new emails were discovered, only to shut it down again ten days later. Why would Comey and McCabe decide that the assistant director should disqualify himself then, but not before? Nothing had changed, except that his hidden conflict of interest had been exposed. Stepping aside in the eleventh hour was an implicit admission that McCabe should never have had any role in the case to begin with. His prior involvement was not only improper and unethical, but seemingly corrupt. Tom Fitton, president of Judicial Watch, which obtained many of the McCabe documents through a Freedom of Information Act lawsuit, said, "These new documents show that the FBI leadership was politicized and compromised in its handling of the Clinton email investigation."[11]

The Justice Department's inspector general opened an investigation into McCabe's failure to recuse himself and whether he violated DOJ regulations in neglecting to properly disclose payments made to his wife.[12] The IG, as well as the Senate Judiciary Committee and the Office of U.S. Special Counsel, also began examining whether McCabe violated the Hatch Act which prohibits certain government officials from campaigning in partisan races. Photographs and other documents suggest he did.[13]

Senator Charles Grassley, chairman of the Judiciary Committee, was unsparing in his criticism of McCabe in a scathing letter to Director Comey on March 28, 2017:

> You have publicly stated that the people at the FBI "don't give a rip about politics." However, the fact is that the Dep-

> uty Director (McCabe) met with Mr. McAuliffe about his wife's run for elected office and she subsequently accepted campaign funding from him. The fact is that the Deputy Director participated in the controversial, high-profile Clinton email investigation even though his wife took money from Mr. McAuliffe. These circumstances undermine public confidence in the FBI's impartiality, and this is one of the reasons that many believe the FBI pulled its punches in the Clinton matter. FBI's senior leadership should never have allowed that appearance of a conflict to undermine the Bureau's important work.[14]

As McCabe's conflict of interest came into sharper focus, Congress demanded answers. He reportedly played a direct hand in altering Comey's statement that originally determined Clinton committed crimes.[15] In so doing, he appears to have been one of three people who helped persuade the director and others to absolve Clinton of any wrongdoing.

After months of resistance and under threat of a contempt of Congress resolution, McCabe finally acquiesced to pressure by agreeing to testify before Congress in December 2017. He came under withering scrutiny during two closed hearings before the House Intelligence Committee and a joint investigation by the House Oversight and Judiciary Committees.[16] His testimony behind closed doors lasted more than sixteen hours. Many members expressed unhappiness with the responses he provided when grilled about whether he allowed his own bias to intrude on the decision to clear Clinton.[17] CNN reported, "The panel's Republicans forced McCabe to answer questions about internal emails they believe showed Comey mishandled the investigation, according to multiple sources. Two Republicans emerged from the Thursday hearing saying McCabe's testimony did not change their belief that Clinton got favorable treatment by the FBI when it decided not to pursue criminal charges last year over the handling of her private email server."[18] What evidence the emails contained was not made public.

Shortly before McCabe's appearances on Capitol Hill, Congressman Trey Gowdy, chairman of the House Oversight Committee, seemed contemptuous of the deputy FBI director's conduct and perceived Clinton bias. He told Fox News, "I'd be a little bit surprised if (McCabe's) still an employee of the FBI this time next week."[19] Grassley was even more outspoken, venturing the opinion that "he ought to be replaced. And I've said that before and I've said it to people who can do it,"[20] a reference to Christopher Wray, who had taken over from Comey when he was fired by President Trump in May 2017.

In a manner, both Gowdy and Grassley were prescient. Two days after McCabe's second acrimonious appearance before Congress, the news broke that he planned to retire within three months when he would become pension eligible.[21] His anticipated exit removed pressure on Wray to demote or fire him for suspected misconduct. A month later, McCabe's name surfaced in a House Intelligence Committee memo as someone suspected of committing government surveillance abuses during the wiretapping of Trump campaign associates. At the same time, the Justice Department's inspector general had accumulated evidence of wrongdoing by McCabe. Shortly thereafter, Christopher Wray removed him from his post as deputy director.[22] McCabe was placed on what was described as "terminal leave" until his planned retirement took effect.

Two days before his planned retirement, McCabe was fired for "just cause." His dismissal was approved by both the FBI and the Justice Department, and endorsed by Attorney General Jeff Sessions after what he described as "an extensive and fair investigation" by the FBI's Office of Responsibility (OPR) and the DOJ's Office of the Inspector General (OIG).[23] Both the OPR and OIG concluded that "McCabe had made unauthorized disclosure to the news media and lacked candor—including under oath—on *multiple* occasions."[24] In other words, McCabe did not tell investigators the truth, more than once, about leaks to the media.

It turns out that the inspector general reportedly determined that McCabe was untruthful four times, twice under oath. According to Representative Jim Jordan (R-Ohio), a member of the House Oversight Committee who obtained copies of the IG report, "It wasn't once, twice, it wasn't even three times. Four times he lied about leaking information to the *Wall Street Journal* about the FBI."[25] The IG subsequently sent to the Justice Department a criminal referral for potential charges against McCabe.

If the findings of the inspector general are true, this would be sufficient justification to bring criminal charges against McCabe for perjury or false statements.[26] McCabe immediately issued a public statement denying he misled investigators and portraying himself as an innocent victim. But at the same time, he implicated Comey in his leak to the media, which could also result in similar charges against the former director. McCabe admitted he shared information with a reporter, but stated the following:

> As deputy director, I was one of only a few people who had the authority to do that. It was not a secret, it took place over several days, and others, including the director (Comey), were aware of the interaction with the reporter.[27]

This seems to stand in stark contradiction to what Comey told the Senate Judiciary Committee on May 3, 2017, when he was questioned at the outset by Chairman Charles Grassley about FBI leaks. Comey denied ever being an anonymous source to the media about the Trump or Clinton investigations. But then he was asked the following:

> **GRASSLEY**: Question two, relatively related, have you ever authorized someone else at the FBI to be an anonymous source in news reports about the Trump investigation or the Clinton investigation?
> **COMEY**: No.[28]

In his statement, McCabe admits he shared information anonymously with a reporter about the Clinton investigation and also contends that Comey knew about it, which means the then-director may have either expressly or implicitly approved the leak. Was McCabe telling the truth or was Comey? It cannot be both.

James Kallstrom, who worked for the FBI for twenty-seven years and became assistant director, said of McCabe, "I think nothing that he said was truthful from my understanding of his role at the FBI. What kind of an investigator was he? What kind of a supervisor? He was lacking in so many ways."[29] Former Assistant deputy director Buck Revell, who knows McCabe, agreed, "I frankly thought he was weak. He didn't have the experience or the backbone to carry out that job. I was not at all satisfied with his abilities to give guidance to the Director or to hold the fort on issues that the senior career person in the bureau should be doing."[30]

In addition to McCabe's conflict of interest involving his wife, there is evidence that, as deputy FBI director, he was instrumental in manipulating the facts and contorting the law to let Hillary Clinton off the hook, notwithstanding compelling evidence she committed a myriad of crimes. It appears he helped sanitize the infamous Clinton exoneration statement read by Comey.[31] At first, it found evidence that she committed crimes, but then the wording was altered to clear her. It is believed that McCabe also signed off on warrant applications to spy on a Trump campaign associate—applications that obscured evidence and potentially deceived the court. This will be examined in forthcoming chapters.

Did McCabe conspire with Comey and others to undermine the Clinton investigation and subvert justice? At the very least, did he allow a personal or political bias to influence the course of a case that should have gone the other way?

If so, there were likely other individuals at the FBI who were involved.

PETER STRZOK AND LISA PAGE

One of McCabe's senior advisers was an FBI lawyer named Lisa Page. She reportedly carried on a romantic affair with Peter Strzok, who served as the FBI Counterespionage Section chief under McCabe. Both Page and Strzok were married to other people, but they worked closely together on the Clinton email investigation as key players involved in much of the decision-making.

It was first estimated that they exchanged some 10,000 text messages during and after the probe, although the figure later ballooned to more than 50,000 messages.[32] Initially, only a small portion of the texts, roughly 375 messages, were turned over to Congress, in December 2017. Immediately, the content of the Strzok-Page communications cast serious doubt on their objectivity and neutrality in assessing whether Clinton committed crimes. The politically charged texts showed a stunning hostility toward Donald Trump, denigrating him as an "idiot" and "loathsome."[33] At the same time, the texts were replete with adoring compliments of Clinton, lauding her nomination and stating: "She just has to win now."[34]

Strzok played a pivotal role in Comey's announcement that he would recommend that no criminal charges be filed by the Justice Department against Clinton. As the lead investigator in the Clinton case, Strzok is believed to have changed the critical wording in Comey's description of Clinton's handling of classified material, substituting the words "extremely careless" for "grossly negligent." According to CNN, citing several sources, electronic records prove it.[35]

Another part of Comey's statement was also changed. Initially, he concluded that the enormous volume of classified documents on Clinton's servers supported the "inference" that she was grossly negligent. This would be sufficient for a finding of criminality. Yet this was also edited out of the final version made public during Comey's announcement on July 5, 2016.[36] Was it Strzok who engineered the edit to exonerate Clinton? Since he made the first edit, it is quite likely he made the second one.

These alterations of key words were integral to the case and had enormous consequences, because they allowed Clinton to evade prosecution under 18 USC 793(f) of the Espionage Act. The changes effectively removed the only legal impediment to her election as president, assuming voters would go along.

The incendiary texts of Strzok and Page were discovered by the Justice Department's inspector general during an internal investigation into Comey's disputed decision to clear Clinton.[37] Impropriety and corruption were being examined.

Here is a sample of several of the vehemently pro-Hillary and anti-Trump conversations contained therein.[38]

PAGE: God Trump is a loathsome human.
STRZOK: Yet he may win.
STRZOK: Good for Hillary.
PAGE: It is.
STRZOK: Would he be a worse president than Cruz?
PAGE: Trump? Yes, I think so
STRZOK: Omg he's an idiot.
PAGE: He's awful.

______________.

STRZOK: God Hillary should win. 100,000,000—0
PAGE: I know

______________.

STRZOK: Turn it on (GOP convention)! The Douchebags are about to come out.
PAGE: And wow, Donald Trump is an enormous do*che.

______________.

STRZOK: Hi. How was Trump, other than a do*che.
PAGE: Trump barely spoke, but the first thing out of his mouth was 'we're going to win sooo big.' The whole thing is like living in a bad dream.
STRZOK: Jesus

______________.

STRZOK: And hey. Congrats on a woman nominated for President in a major party! About damn time! Many, many more returns of the day!!
PAGE: That's cute. Thanks.

______________.

PAGE: She just has to win now. I'm not going to lie, I got a flash of nervousness yesterday about Trump.

______________.

PAGE: Jesus. You should read this (story). Trump should go f himself.

STRZOK: God that's a great article. Thanks for sharing. And F Trump.

______________.

PAGE: God she's (Clinton) an incredibly impressive woman. The Obamas in general, really. While he has certainly made mistakes, I'm proud to have him as my president.

PAGE: We do not want this election stolen from us.

______________.

STRZOK: Trump is a F***ing idiot.

Members of Congress were outraged over the text messages and demanded that more of them be turned over for examination. On January 22, 2018, a second set of 384 pages of texts were made public.[39] One message revealed something stunning: Strzok and Page *knew* that charges would never be filed against Clinton *before* she was ever interviewed by FBI agents, including Strzok. The text dated July 1, 2016, one day before the Clinton interview, made a reference to Attorney General Loretta Lynch's decision to accept whatever conclusion was reached by the FBI:

STRZOK: Timing looks like hell. Will appear to be choreographed. All major news networks literally leading with "AG to accept FBI D's recommendation."

PAGE: Yeah, that is awful timing. Nothing we can do about it.

PAGE: And yes, I think we had some warning of it. I know they sent some statement to Rybicki, because he called Andy.

PAGE: And yeah, it's a real profile in courage, since she knows no charges will be brought.[40]

How is it possible that, in advance of Clinton's interview, a top FBI agent, a key FBI lawyer, and the attorney general would all know that the target of their investigation would not be charged? The answer seems clear. The exoneration of Clinton was predetermined. The investigation was a sham.

How would Lynch know that Clinton would be cleared? On the day he absolved Clinton, July 7, 2016, Comey made a point of saying, "I have not coordinated or reviewed this statement in any way with the Department of Justice or any other part of the government. They do not know what I am about to say."[41]

This statement by Comey appears to have been untrue. According to Page, Lynch knew all about it. Perhaps it was because the attorney general had been personally involved in seeing to it that Clinton was protected. Maybe she had only pretended to exclude herself from deliberations following the public outcry over her infamous tarmac meeting with Bill Clinton days earlier.

A close look at Lynch's recusal shows that it was never a recusal at all. During a televised conversation at the Aspen Ideas Festival on July 1, she announced she would accept the FBI's recommendations.[42] But then, she offered a carefully worded explanation of how her actions were not, technically, a recusal. "A recusal would mean that I wouldn't even be briefed," she said. "While I don't have a role in those findings or coming up with those findings . . . I will be briefed on it and I will be accepting their recommendations."[43]

The question remains: how would Lynch, Strzok, and Page already know that Clinton would skate on criminal charges? Comey claims he didn't decide until after Clinton was interviewed on July 2. Here is what he told the House Judiciary Committee on September 28, 2016:

> All I can do is tell you again—the decision was made after that because I didn't know what was going to happen in that interview.[44]

And yet, Page's text message indicates that the decision not to recommend prosecution was made before Clinton sat down to speak with the FBI, including agent Strzok, and that the attorney general knew all about it. But there is more evidence that Comey made his decision well before Clinton was interviewed.

In a story published by *The Hill,* multiple sources who examined the early draft by Comey said the language of his statement was changed with red line edits to exonerate Clinton on or around June 10, 2016.[45] If true, then Comey appears to have deceived or misled Congress, which is a crime. And the text messages between Strzok and Page serve as corroboration.

With every new text message revealed, the evidence became incontrovertible that the FBI did not conduct a real investigation of Clinton and that her innocence had been predetermined. In a February 25, 2016, message, Page warned Strzok that he should go easy on Clinton because she was likely to win the presidency:

> **PAGE**: She might be our next president. The last thing you need (is) going in there loaded for bear. You think she's going to remember or care that it was more DOJ than FBI?[46]

Additional text messages came to light in early February 2018. In one exchange, dated September 2, 2016, Strzok and Page discussed the preparation of talking points for Comey to give to President Obama, "because POTUS wants to know everything we are doing."[47] At this point in time, the FBI's investigation of Trump-Russian "collusion" was well under way. The messages seemed to put a lie to what Obama had said months earlier when he told Chris Wallace on *Fox News Sunday*, "I do not talk to FBI directors about pending investigations—we have a strict line, and always have maintained it, I guarantee it."[48]

On the day Trump was elected president, Strzok wrote a message, "OMG, this is terrifying."[49] Page replied she was "depressed." A week later she noted that she needed to brush up on "Watergate," and the next day stated, "we have OUR task ahead of us."[50] What exactly she had in mind as their task is unclear, although it may have been a reference to their earlier discussion of an "insurance policy" in the event Trump were to win the presidency.

The bias of Strzok and Page in favor of Clinton and their unabashed enmity toward Trump meant that they should never have been allowed anywhere near the Clinton email case. Despite this prejudice, Strzok was front and center conducting many of the main interviews. He was present on July 2, 2016, during Clinton's surprisingly short questioning by the FBI that was neither recorded nor under oath.[51]

Strzok also interviewed Huma Abedin and Cheryl Mills, two of Clinton's top aides who claimed they never knew about their boss's private, unsecured server until after she left office, even though documents clearly prove that they did, in fact, know.[52] Emails both to and from them show they discussed her unauthorized computer system.[53] However, they were never charged for making materially false statements to the FBI under 18 U.S. C. 1001.[54] When confronted about this by the House Judiciary Committee on September 28, 2016, Comey offered up inane excuses on their behalf, stating that failed recollections should be forgiven in ancillary matters.[55] In truth, there was nothing tangential about top aides who aided and abetted the use of Clinton's unclassified server. Their testimony was central to the case.

Why a specialist in counterintelligence would use his work phone to convey toxic messages to a top FBI lawyer who also happened to be his lover is a case study in hubris and foolishness. But the recklessness with which he voiced his visceral hatred for the Republican presidential candidate and adoration for the Democratic candidate did manage to expose political motives for clearing Clinton and targeting Trump that had long been suspected.

Once Clinton was absolved, it was Strzok who reportedly executed the documents launching the 2016 investigation into Russia's meddling in the American presidential election and whether Trump conspired with Russians to advance his candidacy, despite no real evidence to support the probe.[56]

An editorial in the *New York Post* described it this way:

> As things stand, it now looks like the fix was well and truly in on the Hillary probe. Far worse, it also looks like the "collusion" probe was a partisan hit from the start—which undermines the basis for Mueller's own investigation.[57]

The messages exchanged between Strzok and Page represent incriminating evidence that forces within the FBI appear to have interfered with or influenced their own investigation. If it was done for political purposes, this would constitute obstruction of justice[58] It would also constitute a violation of the Hatch Act.[59] House Judiciary Committee Chairman Bob Goodlatte put it bluntly:

These text messages prove what we all suspected: High-ranking FBI officials involved in the Clinton investigation were personally invested in the outcome of the election and clearly let their strong political opinions cloud their professional judgment.

The taint of politicization should concern all Americans who have pride in fairness of our nation's justice system.[60]

Americans are entitled to a system of justice that is both equal and blind. Those who abuse their authority by imposing their own political prejudices poison this cherished principle of democracy.

CHAPTER 4

CLINTON GREED AND "URANIUM ONE"

There is no greater disaster than greed.

—*The Way of Lao Tzu*, ancient Chinese philosopher

Truth has always been an alien concept to Hillary Clinton.

One of her famous whoppers was the fallacy she tried to sell during an interview with *ABC News* in June 2014. "We came out of the White House not only dead broke, but in debt." [1] It was a ridiculous and gross distortion since Bill and Hillary Clinton already owned two lavishly expensive houses by the time they departed 1600 Pennsylvania Avenue in January 2001.

They had bought a five-bedroom residence in Chappaqua, New York, for $1.7 million and had purchased a seven-bedroom house in the tony enclave of Embassy Row in Washington, D.C., for the hefty price of $2.85 million. People who are dead broke can hardly afford homes valued at a combined $4.55 million.

Regardless, the Clintons felt they were not rich enough, so they wasted little time in fattening their bank accounts and becoming immensely wealthy in a remarkably short period of time. They earned more than $230 million before taxes within a few years of leaving the White House, according to a comprehensive review of their financial worth by *Forbes*.[2]

The source of that staggering amount may be connected to Hillary Clinton's determination to keep her State Department emails forever hidden from public view.

The former president earned much of the cash from speaking fees. At one point, he charged an average rate of $225,000 for each speech, although that figure would sometimes balloon to half a million dollars or $750,000 for a talk that might last no longer than an hour. Income from Bill's books and consulting fees added to the pot of riches, with one client paying as much as $16 million for his business acumen, although he had no prior business experience beyond government employment.[3]

A 2014 analysis by the *Washington Post* determined that Bill was paid roughly $105 million for speeches, with the majority of that income derived from his speaking engagements in foreign countries.[4] Almost half of that money, some $48 million, went into Bill's pocket when his wife served as secretary of state.[5] While paid speeches by former presidents are neither unusual nor prohibited, Bill's reliance on foreign sources demonstrates how the power couple devised a moneymaking machine designed to deliver what foreign governments and corporations wanted most: access and influence.

As Bill Clinton was raking in millions from foreign businesses, investors, and governments, his wife was a senior member of the powerful Senate Armed Services Committee and a member of the Commission on Security and Cooperation in Europe. In those positions, she helped shape decisions that affected governments and businesses overseas. Later, when Hillary became secretary of state, millions of foreign dollars continued to flow to Bill at the very time his wife was America's chief diplomat.

Oddly enough, the pace of Bill's lucrative speaking engagements, especially those abroad, accelerated during the four years his wife presided over the state department. Two-thirds of his fees came from foreign sources. It is no surprise that many of the foreign entities who were shelling out substantial dollars to Bill were the very people and governments who were angling for favorable actions or decisions by Hillary.

In his bestselling book *Clinton Cash: The Untold Story of How and Why Foreign Governments and Businesses Helped Make Bill and Hillary Rich,* author Peter Schweizer described the Clinton profiteering machine this way:

> Who else in American politics would be so audacious as to have one spouse accept money from foreign governments and businesses while the other charted American foreign policy? Or would permit one spouse to conduct sensitive negotiations with foreign entities while in some instances the other collected large speaking fees from some of those same entities?[6]
>
> Meanwhile, bureaucratic or legislative obstacles were mysteriously cleared or approvals granted within the purview of his wife, the powerful secretary of state. Huge donations then flowed into the Clinton Foundation while Bill received enormous speaking fees underwritten by the very businessmen who benefited from these apparent interventions.[7]

As enriched as Bill became, it paled in comparison to the largess heaped upon the Clinton Foundation.

THE CLINTON FOUNDATION GRAVY TRAIN

At the end of the Bill Clinton presidency, the couple established a nonprofit organization named the Clinton Foundation. By all outward appearances, it was a good and legitimate charity. Driven by the celebrity of a former president and the prospect that his wife was positioning herself to make her own run for the White House, the foundation became a magnet for people, businesses, and foreign governments. They calculated that generous financial contributions might reap the kind of rewards that can only be conferred by powerful public officials like the Clintons. It has been estimated that the foundation raised more than a billion dollars over the years.

The charity also became a cash conduit, helping Bill collect millions of dollars as he leveraged the foundation to secure his lucrative personal speaking engagements.

Before she was confirmed as secretary of state, members of Congress expressed deep concern that foreign governments and businesses would seek to buy influence by funneling cash to Bill and even more wealth to the Clinton Foundation. Hillary promised it would not happen and vowed to senators at her confirmation hearing that "all contributors will be disclosed."[8] She insisted she would not only insulate herself from genuine conflicts of interest, but avoid even the appearance of conflicts. She did not keep her promises, not even close. A comprehensive review by *The Hill* found innumerable examples of how Clinton disregarded both her written and oral promises of transparency.[9] Reuters reported that "millions of dollars in new or increased payments from at least seven foreign governments" were never accounted for as promised.[10]

The Clintons were adept at disguising the channeling of cash from donors. Their foundation became a major beneficiary of foreign largess. While operating as a charity, it also served as a ready receptacle for hundreds of millions of dollars in donations from foreign contributors who either directly or indirectly had business before the State Department.[11]

Victor Davis Hanson, a historian at the Hoover Institution at Stanford University, also accused the Clintons of peddling both access and influence in what he characterized as a "shameless" scheme driven by "inordinate greed." In his column "How the Clintons Got Rich Selling Influence While Decrying Greed," he wrote:

> The worth of both the Clinton family and the Clinton Foundation . . . is truly staggering, and to a great extent accrued from non-transparent pay-for-play aggrandizement.
>
> Well before Hillary Clinton's failure in the Democratic primaries in 2008 and her subsequent appointment as secretary of state, the Clintons had found a way to exploit the idea that both of them would return to the White House. That reality gave them access to quid pro quo opportunities often funneled through a philanthropic foundation, of a sort unknown to any past American president.[12]

Many of the largest contributions to the Clinton Foundation were made by people and shell companies connected to a Russian-controlled business called Uranium One, which managed, through a series of clever maneuvers, to seize control of a sizable percentage of America's prized uranium assets.

THE URANIUM ONE SCAM

In September 2005, Bill Clinton traveled to Kazakhstan. Joining him was his friend Frank Giustra, the Canadian investor who wanted to purchase Kazakh uranium mines, which required the government's approval. Upon arrival, Bill met with President Nursultan A. Nazarbayev, who had a long and ugly record of suppressing any political opposition. Several human rights organizations have accused him of severe abuses during his notorious authoritarian rule.[13] Nevertheless, Bill praised Nazarbayev at a public news conference which helped him immeasurably in his quest for improved public relations. Days later Giustra got his lucrative uranium mines. Soon after that, the Clinton Foundation received a $31.3 million donation from Giustra, followed by a pledge to give $100 million more.[14] The deal also provided Bill with incredibly profitable speechmaking fees.

By virtue of a subsequent merger, Giustra's company became a uranium giant called Uranium One. According to the president of the government agency that runs Kazakhstan's uranium industry, Hillary Clinton pressured the foreign government to approve the merger. Clinton, who sat on the powerful Senate Armed Services Committee, had allegedly threatened to withhold U.S. aid if the deal did not go through.[15] It should come as no surprise that it did.

In addition to its ownership of Kazakh mines, Uranium One began to acquire American uranium deposits in Wyoming, Utah, Texas, and South Dakota. Rosatom, the Russian State Atomic Nuclear Agency, which also controls Russia's nuclear arsenal, saw a golden opportunity to enter the valuable U.S. uranium market. It gradually purchased shares of Uranium One, eventually gaining majority control of the company.

By this time, Hillary had moved on to become secretary of state. But Russia's acquisition of a Canadian company that held U.S. uranium deposits had to be approved by Washington since uranium is considered a vital U.S. strategic asset. An executive branch task force called the Committee on Foreign Investment in the United

States (CFIUS) had the authority to accept or reject the Russian takeover of Uranium One. The secretary of state is designated to preside over the nine cabinet officials on the CFIUS. Even though Hillary had, in the past, opposed a foreign government's takeover of an American strategic asset, CFIUS approved the Russian deal.[16] Since any one of the members of CFIUS could have vetoed the matter, unanimity was required. It was approved without dissent on October 22, 2010.

Schweizer put the transaction in perspective:

> The result: Uranium One and half of projected American uranium production were transferred to a private company controlled in turn by the Russian State Nuclear Agency. Strangely enough, when Uranium One requested approval from CFIUS by the federal government, Ian Telfer, a major Clinton Foundation donor, was chairman of the board, a position he continues to hold.[17]

This point bears repeating. Clinton's CFIUS committee approved a Russian takeover of U.S. uranium assets at the same time that the Russian company's chairman was giving substantial sums of money to Clinton's foundation. There is no evidence Clinton ever divulged this blatant, if not potentially corrupt, conflict of interest. The American public had no idea until Schweizer uncovered the information and published it in his book.

Just how much control Russia now has over uranium extraction in the United States has been hotly debated. Some have put it at 20 percent.[18] Regardless, it is a significant amount in the hands of an unfriendly and often hostile adversary on the world stage, given that Russia has a lethal nuclear arsenal and can now tap American resources for the weapons it aims at U.S. cities.

As for the millions of dollars that helped grease the deal, Frank Giustra was not the only one connected to Uranium One who showered the Clinton Foundation with money. As mentioned, Ian Telfer, chairman of Uranium One, made generous personal contributions to the Clinton charity and steered more than $2 million more through a Canadian foundation he controlled called Fernwood Foundation.[19] Again, these donations were never disclosed by Clinton or her foundation as she had promised as a strict condition for Senate approval as secretary of state.

Money also appeared to have made its way to the Clintons from the Russians themselves, although the cash took a circuitous route. A Canadian company called Salida Capital delivered more than $2.6 million to the Clinton Foundation. It just so happens that a company by the same name was listed by Rosatom in its annual report as being wholly owned by the Russian company.[20]

Other people connected to Uranium One made generous donations. All told, some $145 million is estimated to have made its way from shareholders to the Clinton Foundation.[21] Bill Clinton also made out well financially. After Rosatom announced its takeover of Uranium One, but before CFIUS approved the deal, he traveled to Moscow. While there, a Russian financial institution, Renaissance Capital, paid Bill half a million dollars to deliver an hour-long speech.[22] Renaissance had ties to the Kremlin and was also touting Uranium One stock. During his visit to Moscow, Bill met with Russian president Vladimir Putin. No one called this "collusion."

Were all the boatloads of money an exchange of benefits for favorable approval by Clinton and the CFIUS committee vote? Little is known about what occurred behind closed doors. The work of the task force is shrouded in secrecy. Whatever precious few documents may exist are buried somewhere in government archives and have not been made public.

Clinton sought to dismiss allegations of corruption by claiming she was not involved in the uranium decision. Since it was her job to preside over the CFIUS task force, this is hard to believe. Her former assistant secretary, Jose Fernandez, claimed he took Clinton's place on the committee and insisted she never intervened. This, too, seemed implausible since giving a foreign power control over one of our country's most vital strategic assets would normally demand attention at the highest level.

Even more unconvincing was the statement from a Clinton spokesman who maintained that no one had produced any evidence proving that Clinton conferred favors in exchange for money.[23] That's a little like bragging about a burglar who was smart enough to wear gloves. Perhaps the Clintons were adept at hiding evidence. Maybe they avoided any paper trail that would be incriminating. But the timing of contributions and the sheer volume of Clinton enrichment strongly suggest something more than a coincidence. Moreover, the failure of the Clintons to publicly reveal the contributions as promised in their agreement with the Obama White House raises logical and legitimate concerns that they sought to conceal activities that were illicit, if not illegal.

The Russians wasted little time in circumventing their commitment to the United States that Uranium One would not export material beyond America's borders. The Nuclear Regulatory Commission had assured Congress that the Russian-owned company had no license to export the uranium beyond our borders. Cleverly, Uranium One hired a third party to obtain an export license. Uranium yellowcake was then shipped to Canada, which was confirmed by the NRC.[24] From there, some of it was transported overseas with no guarantee that it would not reach, for example, Iran, which was being tutored by the Russians to develop nuclear capabilities.[25] Yellowcake is a concentrated uranium powder that can be enriched and processed for nuclear weapons.

The tangled web of the Clintons, Uranium One, Rosatom, CFIUS, and the tens of millions of dollars in donations to the Clinton Foundation was uncovered by Schweizer and revealed in his book. But other news organizations, in addition to Fox News, examined his evidence and determined that the facts he discovered were accurate and the conclusions he reached were sound.

The *New York Times*, which did extensive reporting of its own, also closely scrutinized Schweizer's book, vetted his sources, and confirmed his key findings in a front-page story of some four thousand words, "Cash Flowed to the Clinton Foundation Amid Russian Uranium Deal."[26] The *Times* reported, "As the Russians gradually assumed control of Uranium One in three separate transactions from 2009 to 2013, Canadian records show that a flow of cash made its way to the Clinton Foundation."[27]

The *Washington Post* did some digging and discovered that 1,110 donors to the Clinton Foundation from a charity closely associated with both Giustra and Telfer at the Russian-owned Uranium One were kept secret and never properly disclosed. Roughly $25 million found its way to the Clinton's Foundation.[28]

Two years later, the *Post* published another story about all the donations from Russia to the Clintons that drew the following conclusion: "It is virtually impossible to view these donations as anything other than an attempt to curry favor with Clinton."[29] But was she influenced in her decision? Clinton has denied it, but the volume and strength of the evidence belie that claim.

RUSSIA USED BRIBES TO SECURE U.S. URANIUM

Before the Obama administration approved the sale of its uranium assets to Moscow in 2010, the FBI discovered that the Russians had secured the deal through an elaborate and illegal conspiracy involving bribery, kickbacks, extortion, money laundering, and racketeering.[30] And yet, it appeared the FBI kept this knowledge secret. There is no evidence that the bureau ever advised key members of Congress, which would surely have stopped the transaction from being consummated. There is also no publicly known evidence as to whether the CFIUS committee itself was told of the Russian corruption before the sale was approved. Americans, most certainly, were kept in the dark.

The evidence of crimes by the Russians was considerable. It included a confidential informant, eyewitness accounts, intercepted emails, secret tape recordings, and financial records. Whether the FBI, a division of the Justice Department, informed Attorney General Eric Holder is unclear. But Holder was among the cabinet secretaries who sat on the CFIUS committee. If he knew and failed to inform fellow members of the committee who approved the deal, he committed an egregious breach of trust.

Did Hillary Clinton know? Unproven. Or covered up. However, in a series of stories broken by John Solomon and Alison Spann of *The Hill,* the undercover informant, later identified as William Douglas Campbell, allegedly witnessed conversations about how "Russian nuclear officials tried to ingratiate themselves with the Clintons."[31] Documents also showed "the transmission of millions of dollars from Russia's nuclear industry to an American entity that had provided assistance to Bill Clinton's foundation."[32]

That entity is APCO, a Washington, D.C., lobby firm, which has numerous former Clinton staffers on its payroll. According to

Campbell, the Russians paid APCO $3 million for one year.[33] The funds were to include free in-kind work for the Clinton Foundation.

Campbell's work as an informant was seen as both valuable and reliable by the U.S. government. So much so, he was paid nearly $200,000, according to the Justice Department.[34] He had intimate knowledge of Russia's illegal scheme to seize control of U.S. uranium, because he worked inside Rosatom posing as a consultant beginning in 2008, some two years before CFIUS sanctioned the sale. His work helped the FBI and Justice Department eventually obtain the convictions of several Russian and American executives. Yet, despite all the incriminating evidence Campbell uncovered for the FBI, Clinton's CFIUS committee approved the sale of America's uranium to Russia. Campbell couldn't believe it; he told Solomon in an on-camera interview:

> I was shocked and appalled because I believed that my reporting to the United States government would head off or would, at least, have some type of consideration before approval was made for those acquisitions.
>
> Because it was a matter of national security, I felt sure that it was being handled on a high level and I was actually assured of that by agents that this strategy would be offset because it was a liability to our country.[35]

Campbell's work exposing Russia's corruption seems to have been hidden or buried as Moscow took control of a substantial amount of America's uranium. While undercover, Campbell also gathered valuable evidence that Russia was assisting Iran's nuclear program of providing equipment and services.[36] The yellowcake uranium that was shipped out of the United States by Russia may well have made its way to Tehran.

Thanks to Campbell, the truth finally emerged. But the damage was done; the sale of America's uranium was consummated. The Senate Judiciary Committee immediately launched an investigation, demanding to know what the agencies that approved the uranium sale knew about the bribery scheme and why the transaction was not stopped in the face of such corruption. They also wanted to know what role Hillary Clinton may have played.

As recently as December 2017, FBI agents interviewed Campbell for five hours about "whether donations to the Clintons charitable empire were used to influence U.S. nuclear policy during the Obama years."[37] Then, in February 2018, Campbell briefed three congressional committees behind closed doors before Congress. He confirmed to them that Moscow routed millions of dollars to an American lobbying firm to influence Clinton.[38] Russian nuclear officials, according to Campbell, expected that some $3 million of their payment to APCO would be redirected to the Clinton's Global Initiative at the same time Rosatom sought approval to purchase Uranium One.[39]

Campbell's testimony to Congress, obtained by *The Hill* and Fox News, described a meeting in which Russian officials "boasted about how weak the U.S. government was in giving away uranium business" to achieve President Putin's "strategic plan" to "take over the uranium industry."[40] Campbell's lawyer, Victoria Toensing, explained it in blunt terms: "(The Russians) were so confident that they told Mr. Campbell that with the Clintons' help, it was a shoo-in to get CFIUS approval."[41]

Did President Obama know? Toensing said the FBI informed her client that the president was aware. "He was told that President Obama had it in his daily briefing twice," said Toensing.[42]

THE LAW

If it can be shown that Clinton used her office as secretary of state to confer benefits to Russia in exchange for millions of dollars in donations to her foundation and cash to her husband, she should be prosecuted under a variety of anti-corruption laws passed by Congress. The most obvious felony is bribing a public official under 18 U.S.C. 201. It makes it a crime if someone in Clinton's position:

> . . . corruptly demands, seeks, receives, accepts, or agrees to receive or accept anything of value personally or for any other person or entity, in return for being influenced in the performance of any official act . . . or being induced to do or omit to do any act in violation of the official duty of such official or person.[43]

This code section criminalizes acts by public officials that are "directly or indirectly" corrupt. There is substantial evidence that Clinton's actions violated this statute.* Based on much of the same evidence, she could also be charged under the federal gratuity statute (18 U.S.C. 201-c),[44] the mail fraud statute (18 U.S.C. 1341),[45] the wire fraud statute (18 U.S.C. 1343),[46] the honest services fraud statute (18 U.S.C. 1346),[47] and the Travel Act (18 U.S.C. 1952).[48]

The FBI evidence, if true, would seem to show that one or more of these illegal "pay-to-play" laws were broken. The government would have to prove that Hillary and Bill got paid, while the Russians got to "play." While prosecutors are required to demonstrate

*Though Clinton's vote was only one of nine, and the other eight all agreed, she would still be liable if her vote was influenced.

a "quid pro quo" or "nexus" (depending on the statute) between the payments and the benefit provided, it appears that the FBI may already possess all the evidence it needs to make a persuasive criminal case.

Moreover, if Clinton leveraged her public office as secretary of state for personal enrichment for herself or her husband while using her charity as a receptacle or conduit for money obtained illegally, it would also constitute money laundering (18 U.S.C. 1956)[49] and racketeering or RICO ("Racketeer Influenced and Corruption Organizations Act," 18 U.S.C 1961-1968).[50]

The Department of Justice defines "racketeering" as the use of a business for a corrupt and illegal enterprise or to operate that business with illegally derived income, domestic or foreign:

> It is unlawful for anyone associated with any enterprise engaged in, or the activities of which affect interstate or foreign commerce, to conduct or participate, directly or indirectly, in the conduct of such enterprise's affairs through a pattern of racketeering activity or collection of unlawful debt.[51]

Thus, repeated acts of foreign bribery involving a U.S. charity would constitute "racketeering activity" under the statute.

The "mafia" and other organized crime syndicates are often prosecuted under the RICO laws. Frequently, they devise a dual-purpose company—one that operates lawfully from the front door, but unlawfully out the back door.

There is little doubt the Clinton Foundation operated as a charity. But if the FBI documents, many of which have not been made public, demonstrate that there was a secondary, hidden purpose devoted to self-dealing and personal enrichment, then Clinton could and should be prosecuted for racketeering.

Finally, in December 2017, amid increasing news reports of corruption and complaints by members of Congress who were demanding a special counsel to investigate Clinton, Attorney General Jeff Sessions ordered federal prosecutors to question FBI agents about the evidence they had uncovered and explain why no action had been taken.[52] Then, in late March 2018, Sessions sent a letter to Congress advising that he had named U.S. Attorney John Huber to conduct a "full, complete, and objective evaluation of these matters in a manner that is consistent with the law and facts."[53] It would be up to Huber to decide whether a special counsel is warranted.

According to the Associated Press, more than half the people outside the government who met with Clinton while she was secretary of state donated money to her foundation.[54] If Clinton was peddling access, was she also peddling influence? Again, the reported FBI documents may hold the answer to that question.

Why has there been no prosecution of Clinton to date? Why did the FBI and, presumably, the Department of Justice during the Obama years keep the evidence secret? Was it concealed to prevent a scandal that would poison Barack Obama's presidency? Was Hillary Clinton being protected in her quest to succeed him?

The answer may lie with the very people who oversaw the Russian bribery investigation and who knew of its explosive impact. Who are they?

MUELLER, COMEY, AND ROSENSTEIN

Robert Mueller was the FBI director during the time of the Russian uranium probe, and so was his successor James Comey who took over in 2013 as the FBI was accumulating even more damaging evidence of Russian corruption. Documents show that Rod

Rosenstein, who was U.S. Attorney for the District of Maryland at the time, was supervising the Russian bribery prosecutions. Andrew Weissmann was then the Chief of the Fraud Section in the Criminal Division at the Justice Department. Both their signatures appear on a plea agreement in the criminal case that was kept quiet. There is no public information that these three men ever advised Congress of any of the incriminating materials they had discovered documenting Russian corruption. It appears they buried the evidence from Congress, key officials in the executive branch, and the American public.

It would seem that these three government officials saw no danger in Clinton's allowing Russia to control more of the world's nuclear power supply, yet they later became the three top people most alarmed by allegations that the Trump campaign conspired with the Russians to influence the 2016 presidential election. Mueller, Comey, Rosenstein, and Weissmann ignored potential crimes involving Clinton and Russia, but were more than eager to conjure some evidence that Trump "colluded" with Russia.

How can anyone have confidence in the outcome of the Trump-Russia matter if the integrity and impartiality of the lead investigators were compromised by ineptitude, or possible malfeasance, in the Clinton-Russia case?

THE "SMOKING GUN" DOUG BAND EMAILS

To understand how Hillary Clinton could be influenced by charitable donations, and want to keep any correspondence about the Foundation hidden, it's important to understand how their scam worked.

In the WikiLeaks emails from October 2016 is correspondence

from Douglas Band who worked for the Clinton Foundation.[55] He described how he not only raised money for the charity but also steered millions of dollars to Bill by using the charity's donors. If the foundation was not operating strictly as a charity under the laws governing nonprofit groups, it should have been deemed an illegal enterprise. In other words, criminal fraud.

Band used his consulting firm, Teneo, to solicit money from corporations for both the Clinton Foundation and Bill personally in what the memo described as "Bill Clinton Inc." Band specifically identified all the sources of the money given by corporate entities to the foundation and then explained how those same companies, at his behest, simultaneously supported Bill:

> Throughout the past almost 11 years since President Clinton left office, I have sought to leverage my activities, including my partner role at Teneo, to support and raise funds for the Foundation. This memorandum strives to set forth how I have endeavored to support the Clinton Foundation and President Clinton personally.[56]
>
> We have dedicated ourselves to helping the President secure and engage in for-profit activities.[57]

How much money did the former president pocket through the combination of for-profit and nonprofit foundation connections? One passage from the Band email is especially revealing:

> Since 2001, President Clinton's business arrangements have yielded more than $30 million for him personally, with $66 million to be paid out over the next nine years should he choose to continue with the current engagements.[58]

A review of the list of donors offered by Band shows that an astonishing amount of cash that went to Bill came from Clinton foundation donors. These are the very same donors who sometimes had business before Hillary's State Department and some of whom appear to have received benefits therefrom.

For example, in 2009 Hillary helped UBS, the Swiss bank, avoid the IRS, and then Bill got paid $1.5 million for speeches. Soon after, their foundation received a tenfold increase in donations. Reporters from the *Wall Street Journal* discovered that "total donations by UBS to the Clinton Foundation grew from less than $60,000 through 2008 to a cumulative total of about $600,000 by the end of 2014."[59] If that was a reward for Hillary's influence, as it appeared to have been, it could constitute the crime of bribery under federal law (18 U.S.C. 201).[60]

It wasn't just UBS that was lining Bill's pocket with cold hard cash. Ericsson and Barclays paid him speaking fees of $900,000 and more than $700,000, respectively.[61] The same companies gave generously to the Clinton Foundation and had overseas business interests over which Hillary could exert influence. More than a dozen other corporations made similar contributions.

When asked to explain what appeared to be blatant double-dealing and the stench of pay-to-play, the Clinton campaign did not really deny it, but simply stated that the charity did wonderful work.

Band wasn't the only person in the Clinton inner circle that was shilling for dollars by using the foundation as a conduit. Huma Abedin, a longtime confidant of Hillary, held three jobs simultaneously: deputy chief of staff to the secretary of state, a paid consultant to the Clinton Foundation, and an income-earning consultant to Band's company, Teneo. Hillary is the one who arranged for the "special government employee" exception to the rules that normally forbid such conflicts of interest. Documents show that Abedin exploited her triple role to grant access to her boss whenever a generous contributor wanted something.[62] The copacetic setup worked well. Donors were accommodated, while the charity enjoyed remuneration.

Band seemed to have known that his pivotal role in the Clinton gravy train was suspect. In an email to Clinton campaign chairman John Podesta, he wrote, "I'm starting to worry that if this story gets out, we are screwed."[63]

Even foundation lawyers knew corruption was likely afoot. After an audit, the law firm authored a draft memo identifying a myriad of problems that could escalate to legal jeopardy. No conflict of interest policy had ever been implemented at the foundation. The December 5, 2011, draft reads like the makings of an indictment:

> In addition, some interviewees reported conflicts of those raising funds or donors, some of whom may have an expectation of quid pro quo benefits in return for gifts.[64]

If a legal team of foundation auditors was able to uncover quid pro quo expectations, imagine what a team of prosecutors would find if they examined in earnest the Clintons and their dual-purpose charity. It is a fair assumption that criminal charges would follow.

THE CLINTON GRAVY TRAIN CRASHES

Once Hillary Clinton lost the presidential election, contributions to the Clinton Foundation and the Clinton Global Initiative began to dwindle.[65] What foreign business or government wanted to shell out big bucks for nothing in return? Hillary was no longer in office, with little prospect for returning to power. She was in no position to confer benefits in exchange for money. No chance for a quid pro quo of favors for cash.

Tax filings showed that donations plummeted 37 percent, while income from speeches dropped from $3.6 million to just $357,500 when Clinton began to distance herself from the foundation before the election and amid reports of possible corruption.[66] Contributions plunged another 42 percent the year she lost.[67] The most recent tax filings won't be available until the end of 2018, but are expected to be even worse.

This is probably the best evidence that corrupt purposes were at the heart of the Clinton Foundation as a vehicle for making money for the Clintons themselves. What is the point of trying to bribe someone who no longer holds the levers of power?

For donors, charitable giving was predicated on gaining both access and influence to a powerful person, Hillary Clinton. When she lost her bid for the presidency, benefactors, especially foreign contributors who had previously shown so much interest in Hillary as senator and secretary of state, began to vanish. As the *Observer* described it, "The Clinton Foundation's downward trajectory ever since Clinton's election loss provides further testimony to claims that the organization was built on greed and the lust for power and wealth—not charity."[68]

CHAPTER 5

THE FRAUDULENT CASE AGAINST DONALD TRUMP

It is the worst oppression, that is done by colour of justice.

—Sir Edward Coke, Lord Chief Justice of England and Wales (The Second Part of the Institute of Laws, 1817)

When those who are entrusted to enforce the law, instead, abuse their power to pursue innocent people in the name of justice it is the worst kind of oppression, as Sir Edward Coke observed.

This is what the FBI did in launching its investigation of Donald Trump. There was never any real evidence of wrongdoing by the Republican nominee for president. There was no reasonable suspicion or evidence sustaining probable cause that those in his campaign were collaborating with Russians to influence the 2016 election.

In its purest form, it was a hoax that was manufactured by unscrupulous high-ranking officials within the FBI and the Department of Justice. Their motives were impure, animated by antipathy for Trump. They were determined to tip the scales of justice and, in the process, undermine electoral democracy.

Armed with nearly unfettered authority, they initiated and advanced a criminal investigation for plainly political reasons. Having engineered a way to spare Hillary Clinton of an indictment, they turned their attention to the only person who could halt her march to the White House—Trump. Perhaps in the back of their minds they believed in the old Stalinist creed, "Show me the man and I'll find you the crime." Surely, there must be something there, they reasoned. The FBI resolved to find it.

Trump didn't know it, but he was about to be framed.

CLINTON CASE ENDS AS TRUMP CASE BEGINS

It is likely no coincidence that almost immediately after Clinton was cleared of any criminal charges, the FBI secretly began its investigation of whether the Russian government attempted to interfere with the presidential election and whether associates of Trump were somehow involved.

A year earlier, in July 2015, hackers had gained access to the server at the Democratic National Committee and Democratic leaders, according to the Office of the director of national intelligence (DNI).[1] Evidence showed that it was the work of the Main Intelligence Directorate, or GRU, Russia's military intelligence agency. After many months of investigation, the DNI issued a report entitled the Intelligence Community Assessment (ICA) in January 2017 that placed responsibility directly at the top of the Kremlin:

> We assess Russian President Vladimir Putin ordered an influence campaign in 2016 aimed at the U.S. presidential election. Russia's goals were to undermine public faith in the US democratic process, denigrate Secretary Clinton, and harm her electability and potential presidency. We further assess Putin and the Russian Government developed a clear preference for President-elect Trump.[2]

The effort to disassemble American elections was an old, but dependable, stratagem. The ICA report points out that such secret activities had been going on for decades dating back to the Cold War when "the Soviet Union used intelligence officers, influence agents, forgeries, and press placements to disparage candidates perceived as hostile to the Kremlin."[3]

Therefore, there was nothing in particular that Trump did that precipitated the Russians to try to disrupt the Clinton campaign. He played no role whatsoever, and there was no evidence in the declassified report that he engaged in any clandestine talks with the Russian leadership or other agents of the state.

Instead, Moscow's objective was to sow discord in the election process wherever possible. This was driven by an historic dislike of democratic institutions in the U.S. political structure and reinforced in the ICA report when it observed, "the Kremlin sought to advance its longstanding desire to undermine the US-led liberal democratic order, the promotion of which Putin and other senior Russian leaders view as a threat to Russia and Putin's regime."[4]

The report did acknowledge that Putin held a "grudge" against Clinton for disparaging comments she once made about him and her "aggressive rhetoric."[5] However, this reference seemed like nothing more than guesswork—both highly speculative and petty. No hard evidence was cited in support. Finally, the report made mention that Putin had indicated a preference for Trump's willingness to work with Russia, especially on matters of counterterrorism in the fight against ISIS. This was consistent with public statements Trump made during the presidential campaign.

Regardless of Russia's desires, the ICA report composed of intelligence work by the National Security Agency, the Central Intelligence Agency, and the Federal Bureau of Investigation found no evidence that Moscow's efforts succeeded in any material way to influence the election. Importantly, despite the entire intelligence gathering by these three agencies over the course of many months, they managed to produce not a shred of evidence of so-called collusion between Russia and the Trump campaign.

However, on its own, the FBI devised a plan to pursue a separate investigation of Trump and his campaign, likely driven by the political aims and animus of top officials at the bureau and like-minded partisans at the Obama administration's Justice Department. Peter Strzok, who played a key role in absolving Clinton, led the FBI's investigation into Trump-Russian "collusion," becoming the deputy assistant director of the Counterintelligence Division. Thus, the agent who expressed such visceral hatred of Trump in text messages to his lover was the same person who reportedly signed the papers opening the investigation of Trump and oversaw every aspect of the case:[6]

> **PAGE**: And maybe you're meant to stay where you are because you're meant to protect the country from that menace. [*This is clearly a reference to Trump, since she provides a link to an article about Trump.*]
>
> **STRZOK**: Thanks. And of course I'll try and approach it that way. I just know it will be tough at times. I can protect our country at many levels, not sure if that helps.

Given Strzok's anti-Trump rhetoric, it is reasonable to conclude that he may already have taken steps to "protect" the country from what he considered would be a dangerous and harmful Trump presidency.

Shortly thereafter, Strzok executed the documents launching the 2016 investigation into Russia's meddling in the American presidential election and whether Trump conspired with Russians to advance his candidacy, despite no real evidence to support the probe.[7] Thus, it appears that two people with strident political agendas and personal bias, Strzok and Page, may have accomplished twin goals of clearing Clinton and accusing Trump, evidence be damned.

Is this what Strzok and Page meant in their texts when they discussed his role as protector of the republic? It makes sense, since Strzok was instrumental in clearing Clinton by rewriting Comey's otherwise incriminating findings. Were Strzok and Page also referring to the investigation of Trump that was commenced right after Clinton was absolved? This too makes sense, inasmuch as Strzok signed the papers initiating the bureau's Trump-Russia probe.

If there is any doubt that Strzok and Page sought to undermine the democratic process, consider this cryptic text, dated August 15, 2016, about their "insurance policy" against the "risk" of a future Trump presidency:

> **STRZOK**: I want to believe the path you threw out for consideration in Andy's office—that there's no way he gets elected—but I'm afraid we can't take that risk. It's like an insurance policy in the unlikely event you die before you're 40 . . .[8]

The reference to "Andy" is believed to be deputy FBI director Andrew McCabe. What was the insurance policy discussed in

Andy's office? It may have been the FBI's investigation of Trump and his associates. Or it might have been the infamous anti-Trump "dossier" that was reportedly used by the FBI and the Justice Department as a basis for a warrant to wiretap and spy on Trump associates. Perhaps it was both. It appears that McCabe was a party to the suspected plot.

When Robert Mueller was later appointed as special counsel to assume control of the FBI's Trump-Russia case, both Strzok and Page immediately joined his team of lawyers and investigators. Eventually, Mueller was informed of the compromising texts on July 27, 2017. According to Assistant Attorney General Stephen Boyd, Mueller "immediately concluded that Mr. Strzok could no longer participate in the investigation and was removed from the team."[9] Instead of being fired, he was reassigned to a position in the FBI's human resources department. Page was also removed from the Mueller probe.

Did Mueller or anyone else notify Congress that the Clinton case might have been corrupted by Strzok and Page or that the Trump-Russia investigation had been seriously jeopardized? It appears to have been covered up. Mueller surely knew that if the truth were revealed, it would further discredit his own probe, which had taken on the stench of spoiled fish.

Several congressional committees could smell it and knew something was amiss surrounding the sudden and unexplained departure of Strzok. Answers were demanded. But the Justice Department and the FBI refused to respond or otherwise produce relevant documents that had been subpoenaed by the committees. It was not until December 19, 2017, the night before Deputy Attorney General Rod Rosenstein was set to appear before the House Judiciary Committee, that 375 text messages, out of roughly ten thousand, were finally produced.[10] Reporters then obtained them.

Members of the House Judiciary Committee believe it was Strzok who plotted to use the unverified anti-Trump "dossier," commissioned by the Democratic National Committee and the Clinton campaign, to spy on Trump associates. During a hearing in December 2017, Representative Jim Jordan asked FBI director Christopher Wray whether Strzok presented the "dossier" as evidence to obtain a Foreign Intelligence Surveillance Act (FISA) warrant from a judge to place wiretaps on members of the Trump campaign:

> **REP. JORDAN**: Did Peter Strzok help produce and present the application to the FISA court to secure a warrant to spy on Americans associated with the Trump campaign?
> **DIRECTOR WRAY**: I'm not prepared to discuss anything about a FISA process in this setting.
> **JORDAN**: We're not talking about what happened in the court, we're talking about what the FBI took to the court. The application. Was Peter Strzok involved in taking that to the court?
> **WRAY**: I'm not going to discuss anything to do with the FISA court applications.
> **JORDAN**: Remember a couple of things, director, about the dossier. The Democratic National Committee and the Clinton campaign—which we now know were one in the same—paid the law firm, who paid Fusion GPS, who paid Christopher Steele, who then paid Russians to put together a report we call the dossier, filled with all kinds of fake news *National Enquirer* garbage.
>
> It has been reported that this dossier was all dressed up by the FBI, taken to the FISA court where it was painted as a legitimate intelligence document, that it became the basis for granting a warrant to spy on Americans.

> I think Peter Strzok, head of counterintelligence at the FBI, Peter Strzok, the guy who ran the FBI's Clinton investigation, did all the interviews, Peter Strzok, the guy who was running the Russia investigation at the FBI, Peter Strzok, Mr. Super Agent at the FBI, I think he is the guy that took the application to the FISA court.
>
> And if this happened, if you have the FBI working with the Democrats' campaign, to take opposition research, dress it all up and turning it into an intelligence document to take it to a FISA court so they can spy on another campaign, if that happened, that is as wrong as it gets.
>
> You could clear it all up, we sent you a letter two days ago. Just release the application, tell us what was in it. Tell us if I'm wrong. But I don't think I am. I think that is exactly what happened, and people who did that need to be held accountable.[11]

If it was Strzok who maneuvered to utilize the largely discredited "dossier," underwritten by Democrats, to investigate Trump and spy on his associates, it would explain his pursuit of an "insurance policy" should Clinton lose the election. He may also have been hiding how the "dossier" was misused. In early 2017, House Intelligence Committee investigators were contacted by an informant advising that "documentary evidence" would show that Strzok was deliberately obstructing the congressional investigation into the "dossier."[12]

Why a specialist in counterintelligence would use his work phone to convey toxic messages to a top FBI lawyer who also happened to be his lover is a case study in hubris and foolishness. But the recklessness with which he voiced his visceral hatred for the Republican presidential candidate and adoration for the Democratic candidate did manage to expose political motives for clearing Clinton and targeting Trump that had long been suspected. Against this backdrop, the persistent claims that Trump "colluded" with the Russians during the election can be traced to Strzok and Page.

Steve Rogers, who served in law enforcement for decades and was a senior naval intelligence officer for the FBI National Joint Terrorism Task Force, believes that Strzok and Page conspired to bring down Trump as a candidate and, failing that, sought to bring down his presidency. "They've done tremendous damage," he said. "They've taken that wonderful, glittering gold shield that people around the world are proud of and they tarnished it badly."[13]

NO LEGAL BASIS TO OPEN TRUMP INVESTIGATION

When the FBI launched its investigation into the Trump campaign, it had no legitimate basis for doing so. Like any government agency, the bureau must abide by a set of strict rules that are set forth in a lengthy book entitled, "The Attorney General's Guidelines for Domestic FBI Operations."[14] The rules contained therein are authorized by statutory law and incorporated into a handbook which agents must study and know. It is called the "Domestic Investigations and Operations Guide" (DIOG).[15]

Under these guidelines, the agency is permitted "to conduct investigations to detect, obtain information about, and prevent and protect against federal crimes."[16] But it must have some *reasonable basis* for commencing an investigation after having first identified "a particular crime or threatened crime."[17] It cannot open an investigation into activity that does not or will not constitute a crime.

There are two kinds of investigations, preliminary and full.

Under the FBI guidelines, "preliminary investigations may be initiated on the basis of any allegation or information indicative of possible criminal activity . . . but more substantial factual predication is required for full investigations."[18] In other words, the FBI is specifically prohibited from searching for evidence of a crime that does not exist under law. There must be "an articulable factual basis for the investigation" that indicates that a crime has or will take place.[19]

Importantly, the same rules that circumscribe the FBI's conduct in *criminal* investigations also apply to *counterintelligence* investigations, although the bureau has somewhat wider latitude in opening a counterintelligence case only because the operations of foreign powers fall under a broader rubric of law called espionage. This allows the FBI to scrutinize any clandestine activities by foreign and domestic actors that might jeopardize America's national security. By its nature, it encompasses greater discretion to collect information and surveil.

However, there must be some reliable *intelligence information* to warrant the opening of a counterintelligence probe. This, the FBI did not appear to have. Devin Nunes, Chairman of the House Intelligence Committee, reviewed "electronic communication" between the FBI and the Department of Justice. Appearing on *Fox Business Network*, he stated that no such intelligence information existed when the counterintelligence probe involving the Trump campaign was initiated. He stated, "We now know that there was no official intelligence that was used to start this investigation."[20]

Obviously, the counterintelligence distinction is susceptible to FBI abuse. Former federal prosecutor Andrew McCarthy in writing about the Trump-Russia probe warned, "The government is not supposed to use its foreign-intelligence collection authority as a pretext to building criminal cases."[21] Yet, this appears to be what the FBI did. Under the guise of a baseless counterintelligence probe, it began investigating the Republican candidate for president.

This invites the inevitable question: what *crime* could Trump and his associates have conceivably committed? Did the FBI abide by the law in making an initial determination of any relevant crime that could have occurred before it opened its investigation of Trump? It appears the bureau did not.

Also, what allegation or information or "articulable" facts did agents like Strzok have in their possession that would lead them to suspect criminal activity was afoot in the Trump campaign by virtue of any direct or tangential contacts it may have had with the Russian government? The answer seems to be that they had little, if any, *reliable* facts or *verified* evidence that would support a crime. Yet, the agency dove head first into a probe of Trump without the required legal justification that the law imposes.

Another provision of law is intended to safeguard American citizens, including political candidates for public office, from overzealous FBI agents:

> These Guidelines do not authorize investigating or collecting or maintaining information on United States persons solely for the purpose of monitoring activities protected by the First Amendment or the lawful exercise of other rights secured by the Constitution or laws of the United States.[22]

As the Republican nominee for president, Trump was lawfully exercising a fundamental right guaranteed under Article II of the Constitution—running for office. Nearly every aspect of his campaign was also protected by the First Amendment's free speech clause. Any attempt by government agents to abridge those rights was a violation of the law, as noted above.

Therefore, the FBI's investigation of Trump, a candidate for president, should have been undertaken only upon a reasonable and good-faith showing that a crime might have been or could be committed and based on reliable, "articulable" facts upon which the bureau became aware. In the absence of credible allegations or information that a specific law was broken, the probe was unlawful.

Periodically, the FBI sought to justify its investigation by claiming it was a counterintelligence probe to gather only information about illicit Russian activities and not a criminal case requiring the bureau to first ascertain a potential crime. This was a subterfuge to hide the true purpose, which was a criminal investigation of Trump and his campaign. As McCarthy explained, "It may be called a 'counterintelligence investigation,' but the objective is to undermine Trump, not Russia."[23]

This became evident when Comey appeared before the House Intelligence Committee on March 20, 2017, and delivered the following carefully prepared statement defining the FBI's objective:

> I have been authorized by the Department of Justice to confirm that the FBI, as a part of our counterintelligence mission, is investigating the Russian government's efforts to interfere in the 2016 presidential election and that includes investigating the nature of any links between individuals associated with the Trump campaign and the Russian government and whether there was any coordination between the campaign and Russia's efforts. As with any counterintelligence investigation, this will also include an assessment of whether any crimes were committed.[24]

The last sentence is the operative one. Comey appears to admit that the FBI's counterintelligence investigation was, in fact, a search for criminal activity involving the Trump campaign. But the FBI is only allowed to open such a probe if it has first identified something that might constitute a crime.

Former Special Assistant U.S. Attorney Fred Tecce believes that the FBI launched an investigation in search of a crime against Trump in violation of the law:

> I have yet to see any evidence of a crime. The way it is supposed to work is that the FBI comes to the U.S. Attorney and says "we believe that this crime is being committed and here are the reasons why." There must first be some evidence of criminal activity. It is not enough to say "we think these guys are doing something nefarious, so let's go investigate them." It's even more troubling in this instance because they used this kind of counterintelligence, which gives them a little more leeway to start an investigation of U.S. citizens. That's an absolute outrage.[25]

So, again, what crimes are we talking about? And what facts did the FBI have in its possession when it decided to investigate Trump over alleged collaboration with the Russians during the campaign? The immutable truth is that there were no crimes and no culpable facts from any credible sources.

"COLLUSION" IS NOT A CRIME

The word "collusion" became a prominent part of American lexicon the moment word leaked out that the FBI was investigating Russia's interference in the presidential election and whether anyone in the Trump campaign was involved.

Although the FBI initiated its probe in July 2016, the bureau kept it hidden from public notice for months. Still, the word "collusion" began to creep into the national conversation when outgoing Senate Minority Leader Harry Reid, a Democrat from Nevada, sent a letter to FBI director James Comey on August 27, 2016. In what was obviously an attempt to influence the election, Reid made his correspondence public—surely hoping it would damage Trump. A key paragraph contained an incendiary claim:

> The evidence of a direct connection between the Russian government and Donald Trump's presidential campaign continues to mount and has led Michael Morrell, the former Acting Central Intelligence Director, to call Trump an "unwitting agent" of Russia and the Kremlin.[26]

Without citing any evidence, Reid had accused Trump of being a Russian agent. He further implied that the Republican nominee's campaign was involved in the cyber theft of emails hacked from Democrats' computer systems. He demanded that Comey investigate, although Reid probably already knew there was a Trump-Russia investigation. His goal was to find a way to make it public.

What precisely prompted Reid's inflammatory accusations? He would not say, but it was later reported that two days before Reid's letter, he had received a visit from CIA director John Brennan, who told the minority leader that "the Russians were backing Trump."[27] Brennan, a well-known partisan who had served in the Obama White House, had every interest in seeing Clinton succeed his boss. The fact that Reid did not mention Brennan but, instead, cited his predecessor, suggests a clever diversion to hide the true source. Documents later obtained by Congress revealed that Brennan's visit to Reid was intended to damage Trump's candidacy and was sanctioned by the White House.[28] Representative Mark Meadows (R-N.C.) later discovered the documents in his work on the House Oversight and Government Reform Committee:

> It appears there was coordination between the White House, CIA, and FBI at the outset of this investigation and it's troubling.[29]

Text messages between the FBI's Peter Strzok and Lisa Page reportedly corroborated the coordination between the agencies and the Obama administration to target Trump.[30] Indeed, one text message dated August 5, 2016 states, "the White House is running this," an apparent reference to the FBI's Trump-Russia "collusion" investigation.[31]

Comey, of course, had already opened an investigation into the Trump campaign. Brennan and Reid likely knew this but were searching for a way to make the matter public before the election. Reid's leaked letter to Comey served the desired purpose. It was picked up by news organizations. Notable was the *Washington Post*'s story which characterized the accusations against the Trump campaign as "collusion."[32] This seemed to have been the genesis of a term that was laden with criminal overtones.

The story, however, gained little traction beyond the initial reporting. So, ten days before the election with the outcome in doubt, Reid fired off another letter to Comey with the goal of contaminating Trump with the same unfounded accusation:

> In my communications with you and other top officials in the national security community, it has become clear that you possess explosive information about close ties and coor-

> dination between Donald Trump, his top advisors, and the Russian government—a foreign interest openly hostile to the United States, which Trump praises at every opportunity.[33]

This time, several news organizations ran with the story. The *Washington Post* called it a "brazen claim . . . even for a man known for bare-knuckle politics."[34] But the newspaper once again described Reid's claim against Trump with the invidious term "collusion." With that, the word seemed to take on a life all its own. Over the next year and beyond, it would be misused and misunderstood in a deliberate effort to falsely accuse Trump of crimes he did not commit. The media, which never bothered to examine the legal meaning of "collusion," embraced it as their cudgel to bludgeon Trump in the press and on the airwaves.

Reid, of course, had a notable history of spreading a lie to damage a Republican candidate for president. In the 2012 campaign, he took to the Senate floor and declared that GOP nominee Mitt Romney had paid no taxes over the past ten years.[35]

Reid had no evidence to substantiate his claim. He made some vague reference to an anonymous "source." In all likelihood, he simply made it up. It was a completely false accusation. Romney's tax returns showed he had, in fact, paid millions of dollars in taxes. Reid's deception earned him Four Pinocchios from "Fact Checker," while "PolitiFact" slapped him with its infamous "Pants on Fire" award.[36] Reid was unrepentant and later seemed to brag about how his spurious allegation may have cost Romney valuable votes.[37]

As for Trump-Russia "collusion," the Office of Director of National Intelligence had found no evidence of any such activity when it made public its declassified ICA report on January 6,

2016, just before the new president's inauguration. That same day, in a private meeting, Comey assured Trump he was not personally under investigation. The FBI director later confirmed this when he testified before Congress in June:

> Prior to the January 6 meeting, I discussed with the FBI's leadership team whether I should be prepared to assure President-Elect Trump that we were not investigating him personally. That was true; we did not have an open counter-intelligence case on him.[38]

While Comey and the FBI had found no evidence that Trump committed any wrongdoing or had otherwise collaborated with Russia, the agency maintained its investigation into whether members of the Trump campaign, *other* than the candidate himself, had "colluded" with Russia. Comey first acknowledged this publicly in March 2016 during his appearance before the House Intelligence Committee.[39]

The revelation that Comey and the FBI continued to pursue "collusion" even after the bureau had participated in the larger and far more encompassing ICA probe, which found no evidence of it, raised serious suspicions that the bureau was chasing a criminal case against Trump for political reasons. These suspicions were reinforced in a March 30 conversation when Comey refused Trump's repeated requests that the director simply reveal the truth to the American public—the president was *not* under investigation for "collusion" with the Russians.[40]

Lost amid all of the speculation about "collusion" was the fundamental fact that no such crime even exists anywhere in American law as it pertains to political campaigns. Comey had tasked his agents with finding a crime that wasn't a crime at all. It was, at its core, a legal impossibility. The FBI could engage in all manner of spectacular jurisprudential gymnastics, but it would never change the fact that colluding with Russia was not, by itself, a violation of any provision in America's criminal codes.

Maybe it should be a crime. Perhaps these events will prompt Congress to consider criminalizing such conduct in political campaigns some day in the future. But other than monetary prohibitions on campaign contributions, there is not a single statute outlawing collaboration with a foreign government in a U.S. presidential election. Or any election, for that matter.

Why, then, were so many people who followed the Trump-Russia saga under the mistaken impression that "collusion" is a crime? Principally, it's because the word is loaded with an historically criminal connotation.

"Collusion" became a prominent part of the legal vocabulary when Benjamin Harrison occupied the White House and Congress passed the Sherman Antitrust Act in 1890 outlawing "collusion" in some business practices.[41] Specifically, price fixing and other anticompetitive activities became a criminal offense under Sections 1 and 2 of the Act. Almost overnight, the word "collusion" was converted into a legal pejorative.

However, "collusion" is only criminal in an antitrust setting. It has nothing whatsoever to do with elections. That did not stop politicians, pundits, and journalists from either misunderstanding the concept and/or misconstruing its application to the Trump-Russia hysteria that reached a deafening pitch.

When I first wrote a column arguing that "collusion" was not a crime in the context of an election, some in the media turned to law professors to try to disprove the argument.[42] While the academics all seemed to concede that "collusion" applies strictly to anti-trust matters, they suggested that other statutes with different names might have been violated if the Trump campaign was conspiring with Russians. Of course, a person can conspire to commit a crime. But what crime?

One professor suggested a fraud statute that makes it a crime to "deprive another of the intangible right of honest services" would be applicable.[43] This, he claimed, would include "fixing a fraudulent election" (which, of course, is redundant). It is true that such a fraud statute exists, and, in the abstract under certain specific circumstances, it might be germane to some other case.[44] But it bears no resemblance to any of the facts or claims in the Trump-Russia case.

This was made clear in *Skilling v. United States* when the Supreme Court ruled that the honest services law only covers bribes or kickbacks schemes.[45] No one has asserted this ever occurred in the investigation of Trump and his campaign, nor is there any evidence of it. So the referenced fraud statute does not apply—not even close.

Another professor pointed to campaign laws that prohibit foreign nationals from donating "money or other thing of value" in a U.S. election.[46] But, again, neither facts nor accusations in the FBI's investigation of Trump support a violation of law on this basis. No one has suggested that the Russians contributed money to the Trump campaign. Even if information was conveyed, it is not considered a "thing of value" under election laws. This will be explored in greater detail later in the book in a discussion of the infamous Trump Tower meeting involving the president's son.

Still another professor referred to a "general anti-coercion federal election law" as a possibility.[47] But this statute is not pertinent either because it forbids acts that "intimidate, threaten, command, or coerce" a government employee involved in political activities and elections.[48] How does that remotely apply to the Trump-Russia case? It does not.

What about the hacking of emails in the accounts of the Democratic National Committee and an associate in the Clinton campaign? To be sure, such conduct is a crime under the Computer Fraud and Abuse Act, as another professor insisted.[49] Indeed, if the Trump campaign conspired to work with the Russians to accomplish this, it would be a conspiracy crime under the Act. But there was never any evidence this occurred. It therefore had no relevance.

It is true that during the campaign Trump once quipped that the Russians should try to locate the thirty thousand emails Clinton destroyed. But a facetious public gibe hardly qualifies as a secret conspiracy. Verbal encouragement to locate missing documents is not enough under the law. Normally, it would require an overt act and agreement to assist in the commission of the crime. It appears that no one outside of the U.S. government, including the Trump campaign, even knew about Russia's hacking efforts until well after they were accomplished and made public two days before the candidate's remark.

The elusive nature of "collusion" did not, however, deter the intensive search for incriminating evidence of coordination between the Russians and Trump or his campaign. Various congressional committees and other agencies spent months scouring records, documents, and written communications. Countless people were interviewed.

After a nine-month investigation, the Senate Intelligence Committee revealed that it had conducted in excess of 100 interviews over 250 hours, held 11 open hearings, produced more than 4,000 pages of transcripts, and had reviewed some 100,000 pages. Every intelligence official who drafted the ICA report on Russian election meddling was interviewed, as were all relevant Obama administration officials. Every Trump campaign official the committee wanted to hear from was interviewed.

What did the committee find? Nothing to speak of—literally. The bipartisan leaders would only say, "The committee continues to look into all evidence to see if there was any hint of collusion." The next day, the Chairman Richard Burr criticized the sensationalized media coverage when he said, "We will find that quite a few news organizations ran stories that were not factual."[50]

It is unlikely Burr's committee had uncovered evidence of Trump-Russia collaboration. He said as much when he informed reporters in his home state, "To date, there has been no evidence of collusion."[51] His co-chair, Senator Mark Warner of Virginia, agreed. When asked on CNN whether there "is any evidence of collusion," he responded, "There is a lot of smoke. We have no smoking gun at this point."[52]

In the months leading up to the Senate committee's update, others with access to the various investigations were equally candid. Leading Democrats, like Diane Feinstein of California and Joe Manchin of West Virginia, confirmed they had seen no evidence of Trump-Russia collaboration. Both sit on the Intelligence Committee. If such evidence existed, they would certainly know about it.

In addition to the Senate investigation, the House Intelligence Committee conducted its own probe, examining some 300,000 documents and interviewing roughly 70 witnesses. In early February 2018, its chairman, Devin Nunes, announced there is "no evidence of collusion."[53] A month later, his committee's majority issued a report announcing the same conclusion.[54] The ranking Democrat on the Committee, Adam Schiff, who had spent more than a year condemning Trump for colluding with the Russians, claimed it "still remains to be seen," but he could offer no evidence after spending months searching in vain.[55] During that time, he had described the evidence as "ample" and "damning."[56] However, two weeks after the committee's report became public, he seemed to concede during a televised interview that there was no proof that Trump or his campaign ever colluded with the Russians to influence the presidential election.[57]

Representative Trey Gowdy, Chairman of the House Oversight Committee which conducted its investigation, also confirmed there was no evidence of "collusion."[58]

The results of these probes put a lie to the repeated statements of Democrats who had consistently condemned Trump in numerous public statements in the preceding months. Senator Harry Reid, a master of false accusations and smears, said the Trump campaign was involved in Russia's hacking of election emails and stated unequivocally, "There is collusion there, clearly."[59] He offered no proof.

Representative Debbie Wasserman Schultz, who was forced to resign as Chairwoman of the Democratic National Committee amid her own scandal, said that contacts between people associated with the Trump campaign and Russians "reeks of collusion."[60] Congresswoman Maxine Waters, known for her inflammatory and overwrought statements, promised she could "guarantee" that Trump was "in collusion with the Russians to undermine our democracy," although she offered nothing in the way of evidence.[61] Neither Wasserman Schultz nor Waters ever produced a scrap of evidence to support their fallacious claims.

Not to be left out, Hillary Clinton implied that Trump "colluded" with the Russians to influence the campaign against her by guiding the hacking of Democratic emails.[62] Her comment was delivered while promoting her book explaining why everyone else caused her to lose the presidential election.

The mainstream media was equally complicit in shaping a misleading and often fictitious narrative that the Trump campaign committed a multitude of crimes by collaborating with Russia. Reports by ABC, CNN, and NBC cited acts of "collusion," but were later retracted or corrected when they turned out to be false.[63]

The intelligence community knew there was scant evidence of any wrongdoing and, when pressed, was forced to admit it. James Clapper, the former director of National Intelligence, twice confirmed that he had seen no evidence of "collusion."[64] As the basis for his conclusion, he cited reports from the National Security Agency, the FBI, and the Central Intelligence Agency. James Comey, when asked if Clapper's assessment was correct, stated, "I think he's right."[65] John Brennan, the former director of the CIA, testified that he didn't know about alleged "collusion"—an improbable claim. But he conceded, "We see contacts and interactions between Russian officials and U.S. persons all the time."[66]

Brennan's predecessor, Michael Morell, who twice served as acting director of the CIA but remained well connected to the intelligence community, was less cagey and coy when he said this:

> On the question of the Trump campaign conspiring with the Russians here, there is smoke, but there is no fire at all. There's no little campfire, there's no little candle. There's no spark. And there's a lot of people looking for it.[67]

The statements by Clapper, Brennan, and Morell are compelling since they were senior Obama administration intelligence officials. As such, they were privy to all of the information gathered by both the FBI and the alphabet soup of intel agencies that were investigating the matter. And, of course, Comey headed the FBI itself which launched the formal Trump-Russia probe. He had access to all of the evidence, or lack thereof.

There are two reasons why the multiple investigations have, to date, failed to produce any evidence of criminal "collusion." First, it is not a cognizable crime under any statute. Other ancillary crimes, as pointed out, have no relevance or application to anything that occurred during the campaign. Second, there are no facts to support this amorphous notion of "collusion." The First Amendment protects the right of Americans to speak freely to, and associate with, any person. Russians are no exception. Candidates are allowed to meet, greet, and exchange ideas or information with foreign nationals during political campaigns.

There was never any evidence that candidate Trump had personal conversations or contacts with Russian government officials or anyone connected with the Kremlin. Indeed, he consistently denied it and no one has come forth to assert that he did. His frustration over being falsely accused of "colluding" with a foreign power during his campaign was understandable. It is why he asked Comey and others to reveal the truth that no evidence existed that would prove otherwise. People who are wrongfully persecuted feel compelled to try to clear themselves of the taint of impropriety. It is the natural instinct in us all. Trump had the same right as a private citizen or public figure or presidential candidate.

TALKING WITH A RUSSIAN IS NOT A CRIME

During the course of the 2016 presidential campaign, associates of the Trump campaign had occasion to meet with or talk to Russians. Some of the encounters were quite brief. None of them constituted criminal activity by Trump or his campaign. Nor did they present a substantial "articulable factual basis for an investigation" by the FBI, as the law demands.

Why? Because it is not a crime to talk with a Russian. There is no statute in the U.S. criminal codes that makes it illegal to have a conversation with, or even receive information from, a Russian. As Americans we are free to travel to just about any country in the world, including Russia; more than 200,000 visit there each year.[68] Ideas and information are routinely exchanged without criminal implications.

We are allowed to engage in business enterprises with Russians. More than one thousand U.S. companies, large and small, do business in Russia.[69] Ford, Boeing, Pfizer, PepsiCo, McDonald's, General Motors, ExxonMobil, Procter & Gamble, General Electric, and just about every major American company have operations there. Hundreds of Western firms have representatives living and working in Moscow and St. Petersburg. In 2016, Russia was the thirty-eighth-largest export market for the United States and the twenty-fourth-largest exporter of goods to America. Our own government does business with Russia, selling more than $100 billion in bonds. In essence, millions of Americans collaborate or, if you must, "collude" with Russians every single day. It is not criminal to do so.

It is not unusual for U.S. political candidates to try to burnish their foreign policy credentials in advance of an election by traveling abroad and meeting with foreign leaders. Ahead of the 2008 presidential election, then-Senator Barack Obama was criticized for his lack of experience in international affairs, so he visited Iraq, Afghanistan, Jordan, Israel, Germany, France and the United Kingdom, meeting with leaders there. While in Germany, Obama met with Chancellor Angela Merkel and Foreign Minister Frank-Walter Steinmeier in a discussion about future American-German relations.[70] No one ever accused Obama of "colluding" with the Germans. He could have traveled to Russia and met with Putin. It would have been perfectly acceptable and legal.

Many other candidates in modern electoral politics have done much the same. Some of them engaged in aggressive actions that would make Trump campaign contacts seem tame by comparison. Senator George McGovern, as the Democratic nominee for president in 1972, attempted to negotiate with the leaders of communist North Vietnam over the issue of prisoners of war.[71] Senator Edward Kennedy allegedly reached out to the Soviet Union on behalf of Democrats in advance of the 1984 election.[72]

Merely traveling to Russia or meeting with a Russian government official on America soil is not, by itself, a violation of any law. These contacts, assuming that any of them were even in the possession of the FBI when the Trump probe began in July 2016, were not sufficient to meet the legal requirement of a "substantial factual predication" in the FBI guidelines.

There were also at least eighteen contacts between the Trump campaign and Russia in the months before the presidential election via emails and telephone calls.[73] But according to a Reuters report, "The people who described the contacts said they had seen no evidence of wrongdoing or collusion between the campaign and Russia in the communications . . ."[74] The conversations were about improving future economic relations and cooperating in the effort to destroy Islamic terrorism. These are the same subjects that candidate Trump spoke about publicly on the campaign trail. There was no odious conspiracy or secret treachery in the making. It was all rather routine in the evolving political landscape.

The Clinton campaign also had contacts with the Russian government. A Kremlin spokesman, Dmitry Peskov, confirmed that

Ambassador Sergei Kislyak met with Clinton advisers because that is part of his job:

> Well, if you look at some people connected with Hillary Clinton during her campaign, you would probably see that he [Kislyak] had lots of meetings of that kind. There are lots of specialists in politology, people working in think tanks advising Hillary or advising people working for Hillary.[75]

But no one in the media made an issue of it. Nor did the FBI investigate the Clintons when Bill traveled to Moscow in June 2010 for a $500,000 speech and a meeting with Vladimir Putin. The reason is simple and obvious. The Clintons and their associates were free to speak with Russians or meet with them, just as anyone in the Trump campaign was entitled to do the same. However, as explained in the last chapter, using a public office to confer a benefit to a foreign government in exchange for money, as Clinton may have done, *would* be illegal.

Yet, the FBI decided to criminalize constitutionally protected speech and actions by Trump and his associates by opening an investigation that the bureau knew would damage the candidate's chances once it was leaked. It might also help to destroy his presidency, if elected. No such comparable investigation of Clinton was opened.

Thanks to a complicit media, several of the Trump associated meetings described above gained exaggerated attention with near hysterical press coverage that implied criminality. But the essential question remained unanswered by the FBI: what legal justification did it have to investigate a presidential nominee when it opened its probe in July 2016 right after it had cleared Clinton in the email case? A close examination of records and reports shows the bureau had almost no reliable facts or evidence required under the law.

Yet Director Comey and his agents pursued Trump and his campaign associates with exceptional zeal in what appeared to be a politically motivated agenda to harm the candidate to the benefit of his opponent. A close examination of the various contacts and conversations with Russians reveals that none of them were consequential or sufficient to launch a criminal investigation.

Yet, none of this deterred the FBI from secretly assigning a government informant to work undercover in an attempt to elicit any incriminating evidence he could find within the Trump campaign by talking with associates of the candidate.[76] When this tactic was uncovered in May of 2018, Trump accused the FBI of spying on his campaign, whereupon the Justice Department announced an investigation to determine "whether there was any impropriety or political motivation" involved and to take "appropriate action."[77]

PAPADOPOULOS'S BAR TALK

George Papadopoulos, a twenty-eight-year-old energy consultant, was selected to serve as a volunteer on a Trump campaign foreign policy advisory panel in March 2016.[78] Though he had no real experience on matters related to Russia, he reportedly felt that improved relations with Moscow would be a good idea for the next president. He pushed for a meeting between Trump and Putin, but the campaign rebuffed the idea.

The next month, while in London, Papadopoulos met with Joseph Mifsud, a Maltese professor, who stated that some Russian officials in Moscow claimed to have "dirt" on Clinton in the form of emails.[79] There appears to be no known evidence that the young volunteer ever passed this information along to the campaign. However, Papadopoulos made mention of the rumored Clinton emails and supposed Russian contacts to an Australian diplomat, Alexander Downer, at a London bar during a night of what was later described as "heavy drinking." [80] Months later, when hacked emails from Democrats were leaked to the public, Downer allegedly contacted the FBI thinking there might be some possible correlation to what he had heard in the bar.[81]

A year and a half later, the *New York Times* ran a front-page story, citing unnamed "American and foreign officials," stating that the Australian diplomat's tip to the FBI is what triggered the bureau's opening of the criminal investigation of Trump-Russia "collusion." [82] Perhaps not coincidentally, the story was published at a time when the FBI was under intense pressure to justify why exactly it had commenced an investigation of Trump to begin with. If the probe was opened based on an unverified "dossier" about Trump that was paid for by the Clinton campaign and the Democratic National Committee, then the investigation itself was highly suspect.

Let's examine the rumor Papadopoulos heard. An unidentified person in Moscow told a professor he had "dirt" on an American candidate. The professor told Papadopoulos, who then may have told an Australian diplomat, who then told the FBI. Forget that the professor denied he ever had such a discussion with Papadopoulos about Russian "dirt" on Clinton.[83] Anyone along the chain of chatter could have lied or exaggerated or mistakenly conveyed the exchange of words. By the time the alleged information was repeated numerous times and eventually reached the FBI, it was what lawyers call "quadruple hearsay." It contained no indicia of reliability and would never be accepted as trustworthy evidence in a court of law. It would be inadmissible in every courtroom in America. Hence, it should never have been used by the FBI as justification for opening a criminal probe of any American citizen, much less a candidate for the presidency. Rumors, innuendo, supposition, and gossip are not a legal basis for criminal investigations. The alleged information must first be verified.

Therefore, the *Times* story made little sense. Papadopoulos was trading grapevine whispers and speculation with no way of knowing whether any of it was true. Nowhere in the article did the reporters raise the essential question of why the bureau would initiate a criminal probe with such scant and legally insufficient evidence.

The *New York Times* reported that Papadopoulos also traded correspondence with a Russian and once met with another Russian. This is neither illegal nor improper. His apparent goal was to improve relations between the two nations by organizing a summit between Trump and Putin. The campaign, however, was not interested in this overly ambitious and premature idea.

Before opening its case, the FBI was required to identify a possible crime. Where was the crime in listening to someone claim that Moscow had "dirt" on Clinton? Where was the crime in repeating that bit of gossip? Even if someone in the Trump campaign acted on the information in some way, it was still not a crime to do so, as will be explained later in the book.

It is more likely that the Papadopoulos conversations were not the impetus for the Trump-Russia "collusion" case, but something else. At best, they served as a propitious pretext for an illegitimate probe. At worst, they were employed as a contrived cover for a politically motivated inquisition by the FBI designed to harm Trump in the campaign or the White House, if elected.

While it was not a crime to speak with Professor Mifsud or pass along his words of conjecture, it was a crime for Papadopoulos not to be forthcoming about it with FBI agents when they approached him in January 2017 for questioning. For reasons that were inexplicable and baffling, he did not tell the truth and was charged with, and pled guilty to, making a "material false statement."[84] He was not charged with "collusion" or any other crime because there was nothing illegal about his contacts or conversations involving Russians. Had Papadopoulos simply told the truth, he would not have been charged with anything.

The *Times* reporters may have been used, wittingly or unwittingly, to deflect attention from the dubious "dossier" and to provide some other justification for the Trump investigation. If this was the goal, it was an implausible and clumsy attempt at misdirection. Under no normal circumstances could a rumor based on multiple hearsays be a proper and legal reason for opening a criminal investigation, especially when the original source was not even identified, and none of the allegations were substantiated.

It's worth looking more closely at the "dossier," and why President Trump's tormentors would rather that we didn't.

CHAPTER 6

THE FABRICATED "DOSSIER" USED AGAINST TRUMP

Apart from the Constitution, the government ought not to use evidence obtained, and only obtainable, by a criminal act.

—Justice Oliver Wendell Holmes on the use of wiretaps (*Olmstead v. U.S.*, 1928, 277 U.S. 438, 469-470)

Everything the Central Intelligence Agency does is cloaked in stealth and secrecy. Subversion is a proficiency. Duplicity is second nature. Against this hushed backdrop, it is not easy to uncover John Brennan's machinations while he was President Obama's CIA chief.

However, a senior aide to the House Intelligence Committee who examined Brennan's actions involving the "dossier" offered this analysis:

> John Brennan did more than anyone to promulgate the dirty dossier. He politicized and effectively weaponized what was false intelligence against Trump.[1]

What exactly did Brennan do? When he learned of the "dossier," he allegedly gave the information contained therein to the FBI and Democrats on Capitol Hill, and certain members of the media were alerted.[2] According to his testimony before the House Intelligence Committee, Brennan seemed to acknowledge that he was among the first individuals to have access to the "dossier" and soon alerted the FBI which then opened its Trump-Russia investigation.[3] He also began telling top members of Congress of his "information indicating that Russia was working to help elect Donald J. Trump president," as reported by the *New York Times*.[4] Whether he told them his "information" came from the unproven "dossier" is unclear, but his access to it would seem to answer that question.

Unanswered is whether Brennan ever took any steps to verify the contents of the "dossier" or affirm its authenticity. As director of the CIA, it was well within his authority to do this since both the author and the alleged sources were foreign actors who supposedly possessed information about illicit activities in the United States. For example, gathering intelligence on Russia's suspected hacking of America's computer systems would be well within the CIA's purview. If Brennan studied the "dossier," he would have inexorably concluded the document was worthless, although that might not have deterred him from using it for political purposes to hurt Trump and help Clinton.

On its face, the "dossier" was a preposterous collection of rumors, innuendos, supposition, and wild speculation. At least one significant part of it contained demonstrably false statements. In its entirety, the set of documents incorporated not a bit of direct evidence. Instead, it was based solely on multiple hearsay accounts from inherently unreliable sources in Russia who were, notably, experts in lies and disinformation. Some sources were apparently paid for passing along the hearsay, which made their claims even more suspect. Although some Russian names were referenced, almost none of the information attributed to them actually came *from* them. The sources were largely anonymous, but their lack of credibility could be determined by the capacious characterizations of their duties and Kremlin connections.

Even more suspect were the financial backers who solicited the injurious material. It was paid for by Democrats and the Clinton campaign. This should have instantly called into question its veracity because of obvious, underlying partisan motives.[5] It was a classic example of what's known as "opposition research," with the notable exception that it appeared to have been entirely contrived. The material was not derived from public documents or voting records as most opposition research originates. Instead, it seemed to have been dreamed up and then propagated by pro-Clinton and anti-Trump forces, notably Brennan among them.

The author of the "dossier" should have been considered inherently untrustworthy, if indeed he was the author at all. As a former British spy, Christopher Steele was trained in deception and chicanery. Reportedly, he never traveled to Russia to gather his so-called "intelligence."[6] Rather, he claimed to have had Kremlin-connected sources, even though he had not set foot in Russia since the early 1990s. Given the change in leadership there during the course of more than two decades, it is highly doubtful that any of his so-called sources held important government positions or otherwise had firsthand access to dependable information.

Upon reading the contents of the "dossier," Brennan must have known that it was no more reliable than a supermarket tabloid. Yet, the evidence suggests that he advocated the use of the dubious document to damage Trump. Stephen Cohen, professor emeritus of Russian Studies at New York University and Princeton, spoke about Brennan's role and what he described as "Russiagate" on a New York radio show, the contents of which were published in *The Nation* under Cohen's name:

> Brennan played a central role in promoting the Russiagate narrative, briefing members of Congress privately and giving President Obama himself a top-secret envelope in early August 2016 that almost certainly contained Steele's dossier.
>
> In short, if these reports and Brennan's own testimony are to be believed, he, not the FBI, was the instigator and godfather of Russiagate. Certainly, his subsequent frequent and vociferous public retelling of the Russiagate allegations against Trump suggest that he played a (and probably the) instigating role. And, it seems, a role in the Steele dossier as well.[7]

James Kallstrom, who once served as assistant director of the FBI, agreed, offering insight gleaned from his conversations with those who had access to Brennan's maneuvers and manipulations behind the scenes:

> My sources tell me (Brennan) was leaking almost weekly or daily, and he was taking that bunch of phony crap supposedly from Russia, and peddling that through the Congress and the media. He was one of the active people. I've known him a long time and I think he's involved, and quite frankly, I think it goes right to the top.[8]

The "top," of course, meant President Obama, who received constant foreign policy briefings from his CIA director. Brennan was an Obama and Clinton loyalist at his core. However, in his role as head of the agency that was dedicated to collecting intelligence, Brennan was supposed to be a nonpartisan and apolitical voice. His actions in proliferating the unfounded "dossier" were plainly political and deeply biased. Author and commentator Paul Sperry interviewed Gene Coyle, a thirty-year CIA veteran who served under Brennan. Coyle described Brennan "as the greatest sycophant in the history of the CIA, and a supporter of Hillary Clinton before the election."[9] He added that "Brennan made it very clear that he was a supporter of candidate Clinton, hoping he would be rewarded with being kept on in her administration."[10]

In his testimony before the House Intelligence Committee in May 2017, Brennan admitted that information he received about the Trump campaign and Russians "served as the basis for the FBI investigation to determine whether such collusion—cooperation—occurred."[11] While Brennan did not specifically refer to the unproven "dossier," it is clear that he took credit for instigating the bureau's investigation of candidate Trump.

Journalist Lee Smith, who writes for the *Weekly Standard* and *Tablet,* described Brennan's role in the Trump-Russia case:

> In other words, the FBI investigation didn't start when the Australians, according to the *New York Times*—or the Brits, according to Brennan's most recent version of the story—contacted the FBI after the Papadopoulos-Downer meeting. No, it started when the director of the CIA decided to start an investigation, when Brennan passed on information and intelligence to the FBI, and signaled the bureau better act on it.[12]

It is true that James Comey, who cleared Clinton, made the ultimate decision to investigate Trump and his campaign. And FBI agent Peter Strzok, who loathed Trump, is the one who signed the papers opening the probe. But it appears to have all been orchestrated by Brennan, whose enmity toward Trump he has not tried to conceal in comments and tweets since election day, 2016. In one message posted more recently, Brennan ranted, "When the full extent of your venality, moral turpitude, and political corruption becomes known, you will take your rightful place as a disgraced demagogue in the dustbin of history."[13] Tired clichés aside, Brennan was anything but apolitical. He was Clinton's advocate and protector. In that role, he politicized phony intelligence and instigated the fraudulent case against Trump.

If there is any doubt that Brennan is capable of such malevolence, consider the ominous warning issued by Obama's United Nation's ambassador Samantha Power, "Not a good idea to piss off John Brennan."[14]

It is reasonable to conclude, therefore, that the Russia collusion hoax began when Brennan seized upon the "dossier" to wreak havoc on the Trump campaign. Precisely when he would have first received it is unclear. However, as explained in the previous chapter, he reportedly met with Senate Majority Leader Harry Reid in August 2016.[15] Two days later, Reid sent the first of two letters to Comey demanding an investigation.[16] Those letters were conveniently made public to sully Trump's candidacy with the taint of scandal in advance of the election.

Brennan wasn't the only top official in the intelligence community that was suspected of peddling the fake "dossier." The director of National Intelligence, James Clapper, was also involved, according to information uncovered by the House Intelligence Committee.[17] Representative Jim Jordan (R-Ohio) described Clapper's role, "The guy who leaked information about that dossier is James

Clapper."[18] Jordan insists Clapper gave the material to CNN where he is now a paid contributor:

> Specifically leaking information I believe, from that January 6th (2017) meeting where they briefed President Trump, then President-elect Trump on the dossier. Someone at CNN got information. We think it was Mr. Clapper who gave it to them. And then a few days later Buzzfeed prints the entire dossier.[19]

Jordan's account is backed up by the "Summary of Findings" published by the House Intelligence Committee when it concluded that Clapper was not forthcoming about media leaks. Finding # 44 states, "Former Director of National Intelligence James Clapper, now a CNN national security analyst, provided inconsistent testimony to the Committee about his contacts with the media, including CNN."[20]

But how did top intelligence officials in the Obama administration get their hands on the "dossier" in the first place? Who commissioned it? Where did it come from?

THE PHONY "DOSSIER"

The origins of the anti-Trump "dossier" began in earnest in April 2016 when Marc Elias, the attorney of record for both the Clinton presidential campaign and the Democratic National Committee, hired a company called Fusion GPS to develop negative information on Trump. Fusion GPS, in turn, hired Christopher Steele, who worked for a private British intelligence firm called Orbis Business Intelligence. Steele had retired as a British agent with MI-6 and had once served as a spy stationed in Russia.[21] Elias's firm was paid $12.6 million during the campaign, while Fusion GPS received $1.02 million, and Steele's company, Orbis, earned $168,000 for his work.[22]

The "dossier" was actually a compendium of seventeen consecutive memos believed to be penned by Steele between June and December 2016. The first memo was entitled "Republican Candidate Donald Trump's Activities in Russia and Compromising Relationship with the Kremlin." It was dated June 20 and alleged that the "Russian regime has been cultivating, supporting and assisting Trump for at least 5 years," but it also accused him of "perverted sexual acts" that could be used to blackmail him.[23] Armed with this implausible document that identified no specific source, Steele decided to approach the FBI with his unverified claims.[24] Remarkably, Steele met in person with the FBI on July 5, the very day that Comey cleared Hillary Clinton in her email scandal.[25] Thus, at the moment the FBI was absolving Clinton, it appears that the bureau was beginning its investigation of Trump into Russian "collusion." Perhaps Steele thought that his document in the hands of the FBI and, later, the media would prevent the Republican candidate from gaining the presidency. This, of course, would surely ingratiate him to his financial benefactors. Shortly after the Steele-FBI meeting, the bureau formally opened its criminal probe of Trump on July 31, 2016, under the pretense of a counterintelligence investigation. In the ensuing months, Steele met with several U.S. news organizations to push his "dossier."[26]

It is obvious why the Clinton campaign wanted to trash Trump with any dirt, however real or imagined, that it could dig up. The expected Democratic nominee was facing a legitimate challenger for the presidency and was determined to stop him at any cost. A million dollars for tabloid garbage may have seemed like a small price to pay given the high stakes of power and the cash machine the Clinton campaign had become. But Steele's motive appears to have been personal. According to evidence produced by the House Intelligence Committee, the ex-spy confided to a Justice Department official that he "was desperate that Donald Trump not get elected and was passionate about him not being president."[27]

Any intelligent person with an ounce of skepticism would have quickly dismissed the contents of the "dossier" as nothing more than a collection of unsubstantiated and far-fetched assertions cobbled together by someone with a motive to smear Trump by inventing the equivalent of sordid fairy tales. It read like a horribly written spy novel. This was the challenge for the Clinton campaign and their allies: how to peddle a fantastic set of lies to the FBI and journalists? It took a while.

The first obstacle, of course, was the "dossier" itself.[28] It is worth reading online if you're looking for a good chuckle. But here are some highlights:

- Russia had been cultivating Trump as a political asset for many years;
- Trump and Russia were exchanging intelligence with each other for eight years;
- Russia had been supporting his candidacy for at least five years;
- Russia had been feeding Trump valuable intelligence on Clinton for years;
- Trump was favored by Russia with lucrative real estate deals;
- Trump's lawyer met secretly in Prague with a Kremlin official;
- Carter Page met secretly in Moscow with two Russian

officials (Sechin and Divyekin), promising to lift Russians sanctions;

- Page and Paul Manafort were intermediaries in the Trump-Russia conspiracy;
- Trump agreed to sideline Russian intervention in Ukraine in exchange for hacked email operation;
- Trump defiled a hotel bed in Moscow where he knew the Obamas had slept;
- Russia was exploiting Trump's "personal obsessions and sexual perversions" for blackmail purposes.

No evidence was offered, and there was never any proof that it was true. Some of it was conspicuously false. All of it was transparently partisan and patently absurd. Yet this document was used by Fusion GPS and Christopher Steele to convince a willing FBI to open an investigation into Trump and spy on his associates. The bureau was either duped or, more likely, became a willing participant in its zeal to steer the election in a way that would conform to its own political preference, namely Clinton.

Paul Roderick Gregory studied the "dossier" within days of it becoming public in January 2017. He is considered an expert on Russia and the Soviet Union, having visited there close to a hundred times dating back to the 1960s and having written several books. Gregory opined that the author's use of words like "trusted compatriots" to identify his anonymous sources, plus other tell-tale signs within the composition, convinced him that the "dossier" was actually compiled by a Russian determined to invent a collection of lies for the purpose of creating mischief:

> There are two possible explanations for the fly-on-the-wall claims of the Orbis report: Either its author (who is not

> Mr. Steele) decided to write fiction, or collected enough gossip to fill a 30-page report, or a combination of the two. The author of the Orbis report has one more advantage: He knew that what he was writing was unverifiable.[29]

Gregory singled out one claim that was particularly ludicrous—that Trump's peripheral adviser on the foreign policy council, Carter Page, promised to lift sanctions against Russia in exchange for a 19 percent stake in the oil company Rosneft, which would amount to a multibillion-dollar bribery scheme to line the pockets of either Page or Trump or both:

> This story is utter nonsense, not worthy of a wacky conspiracy theory of an alien invasion. The huge bribe for (perhaps) lifting the sanctions makes Nikita Khrushchev's hare-brained schemes—for which he was fired—look eminently reasonable.[30]

To its credit, *Newsweek* also turned a critical eye to the "dossier" and published a story entitled "Thirteen Things That Don't Add Up in The Russia-Trump Intelligence Dossier."[31] Besides picking apart the main allegations, the magazine also took notice of what it described as "tortured and borderline non-native syntax" and "basic ignorance of Russia."[32] All of this called into question not only authorship and sourcing, but the veracity of any of the dossier's bizarre and outlandish claims.

In truth, there was little that was intelligent about the so-called "intelligence" document purportedly authored by Steele and/or his company Orbis or a Russian. The only allegation that was subject to verification seemed to be rather easily disproven.

The document alleged that Trump's attorney, Michael Cohen,

secretly met with Russians in Prague in August 2016 to arrange cash payments and devise a cover-up operation. It was provably false. Records show that Cohen was in Los Angeles and New York during the stated time-frame. There are no known international records of him traveling to Prague that have been publicly produced. In reaction to the accusation, Cohen stated, "I'm telling you emphatically that I've not been to Prague, I've never been to Czech (Republic), I've not been to Russia. The story is completely inaccurate, it is fake news meant to malign Mr. Trump."[33]

Lawyers are taught and jurors are instructed that if a witness lies about one part of his testimony, it may be concluded that all of his testimony is untrue. And so it was with the "dossier," but more so. Viewed either through the lens of its individual assertions or in its entirety, it was preposterous. Even a casual reader of the document would reach this conclusion.

And so would a discerning journalist like Bob Woodward, who called it a "garbage document."[34] Frankly, that may be an insult to garbage.

On an edition of *Fox News Sunday*, Woodward, the associate editor of the *Washington Post* who rose to fame uncovering the Watergate scandal, told host Chris Wallace that American intelligence chiefs made a mistake and should apologize.[35]

What did the Russians have to say? Dmitry Peskov, a spokesman for the Kremlin, labeled the unverified memo as "pulp fiction." He said, "I can assure you that the allegations in this funny paper, in this so-called report, they are untrue. They are all fake."[36]

Russian President Vladimir Putin offered that the "dossier" was an attempt by the outgoing American president, Barack Obama, to "undermine the legitimacy of the president-elect."[37] Putin's theory is a pretty good bet. Speaking at a news conference shortly after the document was published, Putin branded as nonsense the notion that Russia had been collaborating with Trump for many years:

> Trump, when he came to Moscow a few years ago, was not a politician. We did not even know about his political ambitions. He was just a businessman, one of the richest men in America. Is someone really thinking that our intelligence agencies are chasing every American billionaire, or what? Of course not. It's just a complete nonsense. [sic][38]

Naturally, the comments by Peskov and Putin should be viewed with skepticism inasmuch as they had the incentive to hide the kind of activity that was alleged in the "dossier." However, Putin's mocking dismissal of the charge that the Kremlin had been cultivating Trump as a political asset for many years and exchanging intelligence with him for eight years was a valid one. If the claim was true, it would suggest uncommon, if not inhuman, prescience on the part of Russian leaders. No crystal ball or set of tea leaves would have envisioned Trump's electoral ascendancy all those years before.

But those in the U.S. intelligence community, the FBI, and Democrats who had access to the "dossier" set aside whatever doubts or misgivings they may have had about the farcical document. Truth be damned, the "dossier" could be exploited for a malevolent objective. That is, it could be weaponized to influence the election. Should that fail, it could always be repurposed to bring down Trump's presidency.

CARTER PAGE AND THE "DOSSIER"

Carter Page played a minor role in the Trump campaign, but became a major instrument in the government's quest to pursue the Trump campaign. As an investment banker and a vice president at Merrill Lynch, Page had been assigned by the company to work in Moscow for three years. Thereafter, he started his own energy investment firm in New York.

At a time when some were questioning the foreign policy bona fides of a businessman turned presidential candidate who had never held public office, the Trump campaign decided it would be prudent to convene a foreign policy advisory council to help deflect the criticism. A board was hastily assembled and Page, by virtue of his background in Russia and energy, as well as his doctorate degree, was selected as a member upon the recommendation of Sam Clovis, who eventually became co-chair of the Trump campaign.

There is no indication the candidate knew Page and, in fact, the "adviser" (if he could even be called that) informed Congress, "I've never spoken with him any time in my life."[39] He stressed that he never communicated with Trump in any form, including texts or emails. Page was not even in attendance at the one meeting the council had in the presence of Trump.[40] The policy council itself appears to have been a device to add some patina to the nominee's credentials on foreign policy where experience was lacking. To suggest that Page was an incidental participant in the campaign is an overstatement. In his testimony before Congress, he described himself as "a junior, unpaid adviser."[41]

Page made a trip to Moscow to speak at a commencement ceremony for the New Economic School. In 2009, President Obama spoke there too, so it may have been the case that the school was simply interested in inviting someone associated, however tangentially, to the upcoming presidential election in 2016.[42] Page alerted the campaign of his speaking engagement and received permission to proceed, although campaign manager Corey Lewandowsky cautioned, "If you'd like to go on your own, not affiliated with the campaign, you know, that's fine."[43]

Page was not a fan of America's foreign policy toward Russia, especially in the Obama era. In this respect, his views were largely consistent with the various public statements made by Trump. Following the speech, Page provided to the campaign what he referred to as a "readout" or synopsis of his visit to Russia.[44] The information contained therein was derived mainly from listening to other speeches, reading Russian newspapers, and watching Russian television. It was not, he insisted, acquired during surreptitious meetings with Russian leaders who were hoping to influence the 2016 election.

According to a transcript of his testimony before the House Intelligence Committee on November 2, 2017, Page suggested to the Trump campaign that the candidate visit Moscow himself to deliver an Obama-like speech on foreign relations. "The idea there was bearing in mind Barack Obama's speech as a candidate in Germany in 2008," said Page. "That was what I was envisioning."[45] The idea was roundly rejected by the campaign.

Page readily admitted to lawmakers that while in Moscow he had several brief encounters with people in attendance who may have been "some senior government officials" connected to the Kremlin.[46] This would not have been unusual, given the university's prominent position among elite Russian institutions. He described his interactions as "greetings and brief conversation."[47] All of them, he said, were benign and inconsequential.

Representative Adam Schiff, the ranking Democrat on the Intelligence Committee, seemed to find something sinister in these incidental encounters, even though Page insisted they were innocuous and did not involve some grand conspiracy to interfere in the presidential election.

Undaunted, Schiff grilled Page about having run into Arkady Dvorkovich, the deputy prime minister of Russia, at the conclusion of the speaking event. According to Page, the two men shook hands and said hello in an exchange that lasted less than ten seconds.[48] Nonetheless, in the congressman's warped view, this constituted some illicit "meeting" as he tried to conflate "met" with "meeting." Page did his level best to explain the difference to Schiff who seemed unwilling to comprehend or accept the difference:

> **CARTER PAGE:** I did not meet with him. I greeted him briefly as he was walking off the stage after his speech.[49]

Perhaps Schiff was gullible enough to believe the contents of the "dossier." More likely, he knew it was worthless but was determined to use it as a political weapon to impugn Trump with a tall tale of corruption and "collusion" with the Russians.

The 208-page transcript of Page's testimony makes for interesting reading. Schiff comes across as someone who believes that any individual with a foreign-sounding name, especially a Russian, must be a nefarious character who is up to no good. Hence, any interaction with such a person necessarily rises to the level of suspected criminality. It does not.

Page is identified several times in the "dossier." On page 9 of the document (the July 19, 2016, memo), it is alleged that Page held "secret meetings" with Igor Sechin, the president of the Russian energy giant Rosneft, and Igor Divyekin, a senior Kremlin internal affairs official. The "dossier" claims the "lifting of western sanctions" were discussed and the Russian officials allegedly "hinted" that they had "kompromat" (compromising information) "on Trump which the latter should bear in mind in his dealings with them."[50] The document notes that "Page was non-committal in response."

As with all of the allegations in the "dossier," this conversation was based on double and even triple hearsay. In other words, someone in Russia supposedly told someone else who allegedly told Steele who then purportedly summarized the information in his "dossier" accurately and without prejudice. Single hearsay is unreliable which is why it is often inadmissible in a court of law. Double hearsay is never allowed. Triple hearsay is a farce.

But beyond the anonymous sourcing obstacles based on rank hearsay, there was never any evidence to corroborate that these conversations with Russians happened. No proof has emerged in the nearly two years the FBI has been searching for it.

Page described the "dossier" as "totally preposterous."[51] The meetings recounted therein never happened, he said. He insisted he did not discuss the Trump campaign with any Russian, never "colluded" with them, and was not involved in the hacking of Democratic emails.[52]

Page 30 (the October 18, 2016, memo) of the "dossier" asserted even more outrageous allegations against Page:

> He (Sechin) offered Page/Trump's associates the brokerage of up to 19 percent (privatized) stake in Rosneft in return for lifting sanctions on Russia. Page had expressed interest and confirmed that were Trump elected U.S. President, then sanctions would be lifted.[53]

Page vociferously denied this event took place, much less a promise to lift sanctions in exchange for what would surely be billions of dollars in a share of Rosneft. "Beyond a shadow of a doubt, there was never any negotiations, or any quid pro quo, or any offer, or any request even, in any way related to sanctions."[54]

It is laughable for anyone to believe that Russia would confer a 19 percent stake in one of its most valuable assets, the state-run oil company Rosneft, in exchange for the prospect of lifting Western sanctions. Based on the asset value and capitalization of the company, the pay-out would approximate $11 billion. Indeed, at the end of 2017, Rosneft executed an agreement to sell 19.5 percent to Glencore, a Swiss company, and Qatar's state-owned wealth fund. The deal was valued at approximately $11 billion.[55]

Why would leaders at the Kremlin offer such an exorbitant amount of money to an American candidate who might not get elected or otherwise carry out a promise to lift sanctions that, by any objective economic model, had not caused anything approaching the equivalent damage to the Russian economy? It makes no sense. If the alleged bribe was intended for Page himself, why would Moscow give such money to a volunteer "unpaid, junior advisor" who had never met the candidate and had few real connections to the campaign? It is a screwball concept that would never pass the "smell test" of even the wackiest conspiracy theorist.

That, of course, did not stop Schiff from advancing the theory. In his opening statement on March 20, 2017, at the committee hearing on Russian interference in the 2016 campaign, the Democratic congressman laid bare his conspiracy theory:

> According to Steele's Russian sources, Page is offered brokerage fees by such (Igor Sechin) on a deal involving a 19 percent share of the company (Rosneft). According to Reuters, the sale of a 19.5 percent share of Rosneft later takes place with unknown purchases and unknown brokerage fees.

> Is it a coincidence that the Russian gas company, Rosneft, sold a 19 percent share after former British intelligence officer Steele was told by Russian sources that Carter Page was offered fees on a deal of just that size?[56]

It turns out that Schiff's claims of "unknown purchases and unknown brokerage fees" were not unknown at all. Glencore and the Qatar fund were the purchasers of the stake.[57] It had nothing to do with, and had no connection to, Page, Trump, or anyone else in the campaign. Glencore, as Page pointed out in his letter to the committee, was founded by none other than Marc Rich, a billionaire commodity trader who was friends with the Clintons and was infamously pardoned by Bill Clinton in the waning hours of his presidency.[58]

As for Schiff's facetious interpretation of a "coincidence," it was nothing of the sort. It was well known in the energy sector and other markets that Rosneft was cash poor and looking for investment infusions outside the country to avoid what Bloomberg called "self-privatization."[59] Indeed, a July 2016 publication reported that a 19.5 percent stake was up for sale "to the large foreign investors."[60] Before that, the company had been shopping around its investor presentation searching for buyers. Thus, Steele's reference to the 19 percent figure would have been known by anyone who was paying attention to Russian energy companies which dominate their economy. It was not a secret. Rosneft desperately needed the $11 billion and the state-run company was in no financial position to give it away as a speculative political bribe. All of this appears to have been lost on Schiff.

Whoever composed the "dossier," whether it was Steele or someone else, learned that Page had traveled to Moscow to give a speech. It was well publicized and even televised within Russia. It would serve as an effective way to smear Trump in the media and tangle him up with an FBI investigation. Perhaps some incriminating evidence could be discovered along the way, but that was only a secondary purpose.

When the FBI learned of the "dossier" in early July 2016, it could have simply questioned Page about its allegations and asked him to account for his whereabouts and conversations while in Moscow. This would have made sense since Page had assisted the bureau once before and became a cooperating witness in prior Russian prosecutions. Indeed, under the FISA law, U.S. law enforcement is required to undertake "normal investigative techniques" before resorting to surveillance, as former federal prosecutor Andrew McCarthy pointed out.[61] The FBI did not comply with the law which speaks volumes about their own political motivations to intercede in the presidential contest.

You'd have to be intolerably ignorant not to recognize the Steele "dossier" for what it was: a politically driven collection of fables designed to defame and discredit Trump. Regardless, once created, it was then appropriated by high government officials at the FBI and DOJ to try to commandeer the election process, defeat Trump, and elevate Clinton. When the odious plot failed to succeed, the conspirators doubled-down and sought ways to destroy the new president.

Spying on Trump was one of their gambits. It was the kind of government abuse of surveillance powers that Justice Holmes argued against.

CHAPTER 7

GOVERNMENT ABUSE OF SURVEILLANCE

We are rapidly entering the age of no privacy, where everyone is open to surveillance at all times; where there are no secrets from government.

—Justice William O. Douglas,
Osborn v. United States, 385 US 323, 341 (1966)

To convince a judge to issue a warrant to search, seize, or surveil someone, the government must demonstrate that its information is reliable. It must present facts it has gathered that come from a credible person. In this case, both the facts and the individual who provided them were untrustworthy. The FBI and DOJ surely knew it.

The Fourth Amendment to the Constitution demands what is called "probable cause" whenever the government desires a warrant from a judge. This means there must exist a *fair probability* that a search (or surveillance) will produce evidence of a crime.[1] The affidavit or application submitted to the court must present sufficiently credible information, normally from witnesses, to establish the probable cause. If the information is *hearsay*, it is vital that the source of the information be a reliable person.

In the FISA warrant application to spy on Page, the FBI knew, early on, that Steele was not a credible source. They learned that his assignment from Fusion GPS was purely for political reasons to damage Trump. They also learned that his efforts were funded by Trump's election opponent, Hillary Clinton's campaign, and the Democratic National Committee. This alone should have been enough for the FBI to disregard Steele and discard his "dossier" as lacking reliability as the law demands.

Then there is the "dossier" itself. As explained previously, it was outlandish in its claims and phony in its composition. Any serious person who bothered to scrutinize its contents would conclude it was bogus on its face. The fact that it was supposedly conveyed from anonymous foreign sources and then purveyed through an alleged series of multiple hearsay conversations made it seem all the more puerile and silly.

The sourcing of the "dossier" was insufficient, by American legal standards, to establish probable cause.

FUSION GPS AND GLENN SIMPSON

The "dossier" may have been sifted from the creative imagination of Christopher Steele, but the endeavor was the brainchild of a former journalist by the name of Glenn Simpson. He, more than anyone else, was responsible for initiating and proliferating the "collusion" myth.

Simpson founded the company Fusion GPS, which conducts investigations and research. Initially, a conservative website hired Fusion to create opposition research on candidate Trump in September 2015. But when it became obvious that the businessman and television impresario would become the GOP nominee, the website stopped funding the work in the spring of 2016.[2]

The Clinton campaign and the DNC, through their lawyer Elias, hired Fusion to develop new negative research on Trump.[3] Only thereafter did Simpson retain the services of Steele and his company, Orbis. Their assignment was to investigate Trump and establish the claim that he was collaborating with Russia. Almost immediately, Steele produced the first in a series of memos that became the infamous "dossier" based on his anonymous, supposed sources.

In testimony before the Senate Judiciary Committee on August 22, 2017, Simpson was repeatedly questioned whether he took any steps to verify the allegations in the document. It appears he did little because, he said, he trusted the ex-spy.[4] Did that mean Steele's information was equally trustworthy? Here, Simpson's answer seemed muddled: "Chris (Steele) deals in a very different kind of information, which is human intelligence, human information. So by its very nature, the question of whether something is accurate isn't really asked."[5]

It *should* have been asked. It was reckless and irresponsible not to ask. Specific and detailed information about the origins and methods of soliciting and securing the materials should have been demanded. The report was so offensive and the accusations against Trump were so severe, that anything less than the strictest scrutiny would have been irresponsible. But it seems that Simpson was more interested in the political havoc he could wreak on the Republican nominee than in accuracy and truth.

One by one, Simpson was questioned about the succession of memos compiled by Steele and whether Fusion GPS ever corroborated any parts of the "dossier." As to one memo, he stated, "Most of this I did not seek to independently verify and was relatively new information."[6] As to another memo, he said, "I can't talk about this as a verification, but I was analyzing this. I analyzed this information in the same manner I analyzed the other stuff."[7] When questioned about still another memo, there was this exchange:

> **Q**: Just moving on to the next memo . . . again, when you take a look at that, was there anything that you independently verified that comes out of this memo?
> **A**: I don't think so.[8]

Simpson gave similar answers as to other memos from Steele. To summarize, he appeared to have done almost nothing to verify the explosive allegations in the "dossier." Beyond its content, what about Steele's anonymous sources? Here is an exchange with Senate investigators addressing that question:

> **Q**: Did you take any steps to try to assess the credibility of (Steele's) sources in the material that he was providing you?
> **A**: Yes, but I'm not going to get into sourcing information.[9]

Perhaps he declined to talk about it because there was no "sourcing information." It is hard to imagine that Simpson, who did not speak Russian and did not travel to the country while developing his anti-Trump opposition research, would have been in a position to verify Steele's so-called sources, especially since they were anonymous. He admitted that none of his own work made it into the "dossier," which suggests he did little except pay Steele for his tall tale of Trump's "collusion" with the Kremlin.

A full reading of Simpson's 311 pages of testimony will leave you with the distinct impression that he may never have asked whether

Steele's information was accurate. Moreover, there is no indication Simpson or anyone at Fusion GPS searched for any independent proof to substantiate the many scurrilous claims contained therein.

That did not, however, stop Simpson from conveying the information to journalists. He and Steele shared with reporters the contents of the "dossier" in the summer and fall of 2016,[10] yet he conceded he did not tell them it was opposition research funded by the Clinton campaign and Democrats which would surely have affected the media's view of it. He was asked this question about providing material to journalists:

Q: And they're aware you're being paid to do that research for a client?

A: I don't know. Generally, that's not an issue.

(Brief exchange with Simpson's lawyer)

Q: My question was whether or not they knew you were being paid to do that research.

LAWYER: He answered that question too, and he said he did not explain that to the journalists.[11]

Other than his disdain for Trump's "suitability" to be president,[12] Simpson was convinced there was some illicit collaboration between the Russians and the campaign. He seemed to have formed the opinion that Carter Page's speech in Moscow constituted a sordid element of the conspiracy,[13] even though it was perfectly legal and fairly innocuous. The fact that Page had earlier been caught up as a witness in a Russian spy-ring case appeared to have reinforced Simpson's impression of potential criminal conduct.[14] As mentioned previously, Page was never charged with any wrongdoing and appears to have been a target of a Russia operative, not a willing participant in a spy ring.

Although Simpson had no real proof of Trump-Russian collaboration, he testified that he thought Donald Trump Jr.'s meeting with a Russian lawyer in Trump Tower in June 2016 was a form of corroboration, even though no one in the media discovered the meeting for at least another year.

After Simpson's experience before the Senate committee, the House Intelligence Committee subpoenaed two other executives of Fusion GPS, Peter Fritsch and Thomas Catan. As to every question posed, the two men invoked their Fifth Amendment rights against self-incrimination. They likely knew, better than Simpson, the extent to which they were facing legal jeopardy.

Fusion GPS has long been accused of employing underhanded tactics. In advance of Simpson's appearance, the Senate committee heard sworn testimony from others who claimed that they, too, had been victimized by "prepared dossiers containing false information" and "carefully placed slanderous news items."[15]

The anti-Trump "dossier" was another quintessential operation run by Fusion GPS. Kimberley Strassel of the *Wall Street Journal* wrote, "Fusion is known as a ruthless firm that excels in smear jobs."[16] But the Trump assignment was their most notorious. The title of Strassel's column, "The Fusion Collusion," was felicitous. She took aim not just at the firm, but at the Democratic Party, "whose most powerful members have made protecting Fusion's secrets their highest priority."[17] It was "collusion" at an atrocious level.

Under questioning, Simpson confirmed that Steele made contact with the FBI during the first week of July just as the bureau was absolving Clinton.[18] Simpson admitted he agreed with Steele's decision to reach out to the bureau and probably encouraged him to do so. Simpson and Steele both spoke with the media to push their "dossier" on them because, as Simpson stated, "some of it was gathered for the possibility that it might be useful to the press."[19]

How hard did Simpson push? Perhaps to the point of actually *paying* journalists. On November 21, 2017, the *Washington Examiner* reported on the existence of court documents that showed Fusion GPS made payments to three journalists during the same time Simpson was talking to reporters about the "dossier."[20] The names of the journalists were withheld, but the three "are known to have reported on Russia issues."[21] Fusion's lawyer argued that the payments were compensation for information from the journalists, not to "pay journalists to write stories."[22] It was an excuse that shattered common sense.

While Simpson and Steele were talking to reporters from five different media outlets and Steele was meeting with the FBI, Fusion GPS and Steele were also promoting the "dossier" to a high official at the Justice Department, Bruce Ohr. It turns out that his wife, Nellie H. Ohr, was a paid employee of Fusion GPS who wrote extensively about Russian topics. According to the House Intelligence Committee, she provided "opposition research" on Trump in collaboration with Steele's "dossier."[23] Her husband, who was associate deputy attorney general, held secret meetings with both Simpson and Steele.[24] The Ohr-Fusion entente was not disclosed by Simpson when he testified before the Senate Judiciary Committee.

Nor was it disclosed by the Justice Department when it learned of Ohr's meetings with the proponents of the "dossier" and quietly demoted him.[25] It kept the scandal hidden, perhaps in the hopes that no one would discover that Simpson and Steele were covertly providing the DOJ with the unverified document, funded by the Clinton campaign, that was being used to spy on her opponent's team. Even after the November presidential election, Ohr was meeting with Simpson as the secret surveillance of Page continued.[26] It is not clear what actions Ohr, armed with the inflammatory "dossier," may have taken because the Justice Department has never been forthcoming about what exactly he did.

The DOJ's efforts to cover up Ohr's activities were unconscionable. And, so too, was Ohr's conduct. Since his wife worked for Fusion GPS and contributed to the "dossier," the relationship presented a disqualifying conflict of interest for Ohr who was legally obligated under DOJ regulations to recuse himself from any investigation in which his wife was involved. He did not seek a waiver of the conflict. Upon joining the Justice Department, he had signed an agreement stating that he would be fired for violating its rules. Inexplicably, he was not terminated, which only reinforces the impression that impropriety and concealment continued at the highest levels of the department.

Not only did Bruce Ohr fail to disclose that Fusion GPS was paying his wife, but it appears he did not fully report the nature of the work performed in financial disclosure reports as required under Justice Department regulations.[27] Willfully filing a false report constitutes a crime under 18 U.S.C. 1001.[28]

Hillary Clinton tried to dismiss her role in paying for the "dossier" by stating that it was simply "opposition research."[29] While it is true that political campaigns often use such research, what Clinton, Simpson, Fusion GPS, and Steele did was antithetical to any definition of "research." In a column published by *The Hill*, Ned Ryan explained it this way:

> Opposition research is based on fact, from voting records, court records and public statements, to tax returns and business relationships. Fusion GPS's dossier, on the other hand, was misinformation. It was not opposition research because it was not based on fact.

> Given Fusion GPS's dependence on Russian gossip spread by Vladimir Putin's spies, there is a good case to be made that Fusion GPS more deeply colluded with the Russians than anyone else.[30]

And so, too, did Clinton since she helped fund the effort.

There was never any evidence Trump or his associates "colluded" with Russia to win the election. But there is substantial evidence that Clinton and Democrats "colluded" with Simpson, Fusion GPS, and Christopher Steele, a foreign national, to influence the election by defeating Trump. And Russians, according to Steele, participated by providing real or imagined Kremlin "sources."

Distilled to its essence, the scam worked like this: Clinton's campaign paid Simpson who paid Steele who allegedly got information from Moscow intended to damage Trump and help Clinton. Then the FBI and the Justice Department exploited the dubious information as a pretense to open an investigation of Trump, while the media ran with their stories intended to bring down the new president. The government abused its powers and the press was shamefully complicit.

The two congressional intelligence memos, when read in light of Steele's admissions and Simpson's testimony, give sustenance to the inescapable conclusion that the "dossier" was a fraud. Its author was considered by the FBI to be a liar. Despite this, government officials at the FBI and DOJ secretly collaborated with both Steele and Simpson to obtain their warrants to spy on the Trump campaign without full disclosure of the facts and the kind of legal justification required in our system of justice. When Trump was elected, and even after he took office, the surveillance continued.

The chronology of events demonstrates that the fictitious "dossier" was the pretext for the FBI's probe of Trump, not Papadopoulos's so-called "bar talk." Steele composed his first memo alleging Trump-Russia "collusion" on June 20, then met with the FBI on July 5. On July 31, the investigation of Trump was formally opened. Only thereafter, on August 2, did FBI agents meet with the Australian diplomat to gather information about Papadopoulos.

THE SPY GAME

Carter Page was never charged with any crime. The reason is quite simple: he did nothing wrong. Instead, he became an unwitting pawn in the government's attempt to implicate Trump for something—anything.

Page cooperated with the FBI beginning in 2013 when the bureau was investigating three Russians suspected of trying to recruit an American to act as a foreign agent. It was understandable that Page had communicated with numerous Russians, given his global energy interests and the three years he spent in Moscow. He had no idea that one of the people with whom he had spoken was a Russian operative in a suspected spy ring. The FBI reportedly surveilled Page through a succession of Foreign Intelligence Surveillance Act (FISA) warrants beginning in 2014, but found no evidence to implicate him and, thus, no charges were brought against him.[31] Instead, Page is on record stating that he actively assisted the bureau in its prosecution of the case.[32] All known evidence indicates this is true. He was working with the FBI against Russia, not for it.

Under the FISA law, "recruitment" as a foreign agent is not enough to sustain the burden of probable cause to surveil someone. The statute is quite strict and limited in its scope. Former federal prosecutor Andrew McCarthy explained this quite well in a column he penned for *National Review*:

> The fact that a foreign power is trying to recruit an American to become an agent for that foreign power is not a sufficient basis to issue a surveillance warrant against the American under FISA. It would, of course, be sufficient to issue a warrant against the foreign spies who are making the recruitment efforts, but it is not enough for a warrant against the American citizen who is the target of the recruitment effort.
>
> To get a surveillance warrant under FISA . . . the FBI and Justice Department must establish probable cause that the person to be monitored under the warrant is acting as an active, purposeful agent of a foreign power—not that the foreign power hopes to turn him into such an agent.[33]

Fast forward to 2016; the FBI learned that Page had been named to the Trump foreign policy council. Newly armed with the "dossier" that identified Page as participating in a fictitious meeting with Russians while delivering his speech in Moscow, bureau agents had the perfect excuse to begin spying on Trump and his campaign by seeking a FISA warrant to wiretap Page's phones in October and gain access to his stored communications.

There were legal impediments to convincing the Foreign Intelligence Surveillance Court (FISC) to grant another warrant to spy on Page. New evidence would have to be provided that would justify yet another wiretap of a person who had been tapped before with no result.

A majority of the House Intelligence Committee later concluded that the endeavor required sleight-of-hand trickery by the FBI and the Justice Department to gain the court's permission.

Evidence would have to be concealed, and the judge would have to be deceived. All that was necessary was to inflate the importance of the "dossier," obscure the true nature of its authorship, downplay its fabricated content, camouflage its partisan motives, and hide that it was paid for by the true target's political opponents.

HOUSE INTELLIGENCE COMMITTEE MEMO

On February 2, 2018, the House Intelligence Committee declassified a four-page memorandum and made it public.[34] The findings composed by a majority of Republicans on the committee were derived from testimony under oath and documents obtained pursuant to subpoenas. Here were the main points of what became known as the "Nunes Memo," named after the chairman of the committee, Congressman Devin Nunes:

- FBI deputy director Andrew McCabe testified "that no surveillance warrant would have been sought from the FISC without the Steele dossier information";
- Christopher Steele compiled the dossier and admitted to Deputy Attorney General Bruce Ohr that he "was desperate that Donald Trump not get elected and was passionate about him not being president";
- Steele's bias was recorded in FBI files but not reflected in any of the warrant applications presented to the court;
- Ohr's wife worked for Fusion GPS and assisted in the anti-Trump research, and Ohr gave his wife's research to the FBI;
- The Ohrs' relationship with Steele and Fusion GPS was concealed from the court;

- The FBI assessed the dossier "as only minimally corroborated";
- The dossier was funded by the Hillary Clinton campaign and the Democratic National Committee (DNC);
- None of the surveillance warrant applications disclosed this funding, "even though the political origins . . . were then known to senior DOJ and FBI officials";
- The FBI and DOJ cited a Yahoo News story about Page to support the warrant application, without disclosing that it came from the same source as the dossier—Steele;
- Steele was terminated by the FBI because he "lied" to the bureau about his contacts with the media and disclosure of information;
- After Steele was fired, he continued contact with Ohr at the DOJ;
- FBI director James Comey briefed President-Elect Trump on the dossier at a time when he knew it was "salacious and unverified";
- Comey, McCabe, and others signed the warrant applications when they knew it was unreliable and unverified.

The results of the committee's investigation were stunning. The FBI and DOJ *knew* the "dossier" was unverified conjecture. They *knew* it was paid for by Clinton and Democrats and was politically motivated. They *knew* the person who compiled it despised Trump and harbored a clear bias. And they *knew* they could not spy on a Trump associate without the "dossier." Despite all of this, the FBI and DOJ used it anyway, deceiving the court by actively withholding or obscuring every bit of vital information.

The government spied on Carter Page by obtaining four different warrants at ninety-day intervals, which means the surveillance lasted for roughly a year starting October 21, 2016. But it also allowed the FBI to get its hands on any of Page's stored communications he had in the past. It was a way to spy on him going forward and backward. Nothing of any consequence was ever found, which only reinforced how unwarranted the intrusion was to begin with.

The House Intelligence Committee's evisceration of the original FISA application demonstrates that the judge who approved the initial warrant, had he known the truth, would surely have rejected the government's request to spy on Page because the application was based substantially on faulty evidence submitted to the court and FISA procedures were ignored. While the basis for the subsequent renewal applications for the warrant have not been publicly disclosed, we can reasonably assume they were riddled with the same defects and deceptions. Among those who signed the warrant applications were then-FBI Director James Comey and current Deputy Attorney General Rod Rosenstein.

It did not end there. When Trump fired Comey, it was Rosenstein who appointed Comey's good friend and former colleague, Robert Mueller, as Special Counsel to investigate Trump. These all-too-cozy relationships, solidified amid arguably corrupt acts, served as the underpinnings of the case against the president. Such egregious conduct by high government officials was anathema to fairness and justice.

DEMOCRATS' "REBUTTAL" MEMO

Predictably, Democrats who had fought so hard to stop the release of the Intelligence Committee memo condemned its findings even before it was made public. House Minority Leader Nancy Pelosi called it "bogus."[35] She and Senate Minority Leader Chuck

Schumer demanded that Devin Nunes be removed as chairman of the committee.[36] Representative Adam Schiff claimed the memo was "rife with factual inaccuracies" and gave a "distorted view of the FBI."[37] Hardly.

The "Nunes Memo" was neither bogus nor inaccurate. There was no discernable distortion. Each and every finding was based on the sworn testimony of witnesses who had been interviewed or documents that were obtained by the committee, many pursuant to lawful subpoenas. The memo was short on analysis, but long on established facts. None of the released information jeopardized sources or methods in intelligence operations. Nonetheless, the FBI took the unusual step of issuing a statement strongly objecting to it being made public.[38] Such criticism by the bureau was understandable, given that the information detailed how the FBI and DOJ had seemingly abused their powers to spy on a former Trump campaign adviser.

Immediately after its release, Schiff vowed to construct his own memorandum to serve as a "rebuttal." He and his colleagues on the committee did so. But unlike Democrats who had voted to stop the Republican memo's release, Republicans voted unanimously to make public the Democrats' memo.[39]

While Schiff had suggested that his memo would contradict many of the findings of the "Nunes Memo," it accomplished precious little of that. Instead, a close reading of the document showed that it largely supported the earlier findings, although the new memo strained to massage the facts in a light most favorable to the FBI and Justice Department. It also strove to deflect attention to other issues extraneous to the warrant application. On balance, Schiff's memo contained mostly opinions and argument in a pronounced attempt to reimagine what had so obviously occurred before the FISA court. Conclusions were drawn in the absence of foundational facts.

For example, the new memo claimed that the government

had followed all the proper procedures before the court, "did not 'abuse' the FISA process," and "met the rigor, transparency, and evidentiary basis needed to meet the FISA's probable cause requirement" for the issuance of a warrant to surveil Page.[40] These statements were not even close to being accurate. As explained earlier, the FBI did not interview Page when it learned of the "dossier" even though it was required to exhaust all "normal investigative techniques" before seeking a warrant to surveil.[41] Page had previously been a cooperating informant for the FBI. Instead of spying on him, the bureau should have simply talked to him to determine whether the allegations against him bore any truth and whether the surveillance was even merited.

Schiff's memo asserted that the FISA court was properly informed about the partisan backing of the "dossier."[42] This was sincere in only the most distorted sense. Burying cryptic references in a document's footnotes that disguise the specific names or identities of the Clinton campaign and the DNC hardly constitutes an informed court, especially when the funding of the "dossier" by a political opponent is an indispensable fact for consideration by a judge who is attempting to discern both credibility and motivations. Steele's partisan funding should have been clearly and deliberately highlighted for the court. Rather, it was intentionally camouflaged.

The Democratic memo went to extraordinary lengths to argue that Steele was "credible," ignoring the fact that the FBI fired him for lying and was relying on mostly anonymous sources whose credibility, much less identity, could not be determined or substantiated. This was not fully or truthfully explained to the court. A principal source who lies has no credibility, which means his claims utilizing other unidentified "sources" who supposedly gave him information are equally untenable.

The most astonishing claim in Schiff's memo was that, after the first surveillance, "DOJ provided additional information obtained through multiple independent sources that corroborated Steele's reporting."[43] Conveniently, this information was heavily redacted so there was no way of knowing whether it was true. Common sense dictates that it was not. If support for Steele's claims was collected during the surveillance or through other means, why wasn't Page ever charged with a crime? The answer should be obvious. There was no corroborating evidence. In the alternative, whatever was collected was of insignificant evidentiary value.

Finally, most notable in the Democratic "rebuttal" memo was what was missing. When the "Nunes Memo" revealed that deputy FBI director McCabe testified that no warrant would have been sought without the "dossier," Democrats insisted this was erroneous and that their memo would dispute this finding.[44] It did not. Why? It was omitted likely because what McCabe reportedly said to the committee was true.

Fred Fleitz, who spent nineteen years at the CIA and also once served as a senior staff member on the House Intelligence Committee, described it this way:

> This is one of the most crucial points in the Republican memo, but the Democrats ignored it.
>
> This is a major omission. Either McCabe did or did not say the Steele dossier was the crucial information that drove the FISA warrant request. The committee's Republicans should declassify the transcript of McCabe's remarks to expose what is very likely a huge Democratic misrepresentation.[45]

In the end, the Democratic "rebuttal" managed to rebut almost nothing of any consequence. How could it? Director Comey admitted to Trump that some, if not all, of the "dossier" was "salacious and unverified."[46] Reportedly, Deputy Director McCabe reiterated the same to Congress when he testified behind closed doors.[47] Most importantly, Page had been a credible witness for the FBI, and there was no trustworthy or corroborating evidence he had done anything wrong. None of this deterred the government from relying on the "dossier" to spy on Page in an apparent attempt to inflict harm on Trump, his campaign, and his presidency.

The evidence shows that the bulk of the warrant application was wholly contingent on Steele and his dirty "dossier." There was nothing else. Without it, the FISA court would never have countenanced surveillance by granting the initial and subsequent warrants. Substantial reliance on the "dossier" was confirmed by yet another committee's findings on the other side of the Capitol.

THE GRASSLEY-GRAHAM SENATE MEMO

It wasn't just the House Intelligence Committee's majority members who found shocking evidence that government officials had deceived the FISA court to spy on Trump campaign adviser Carter Page. On the Senate side, the Judiciary Committee conducted a separate investigation and gathered its own set of incriminating evidence.

A memorandum co-signed by Committee Chairman Charles E. Grassley and Chairman of the Subcommittee on Crime and Terrorism Lindsey O. Graham set forth an equally compelling case of wrongdoing by the FBI and Department of Justice in obtaining their surveillance warrant which allowed them to secretly listen in on Page's conversations and to seize any of his past electronic communications such as text messages and emails.[48] Bear in mind, the warrant was obtained *after* Page had left the Trump campaign in the fall of 2016.

The Senate memo completely confirmed the findings contained in the House memo, but produced even more evidence of impropriety by high government officials:

- The FBI never corroborated any of the main claims in the dossier;
- The FBI represented to the FISA court that Steele was "reliable" and "credible";
- The FBI learned Steele was not reliable and had "lied" almost immediately after seeking the initial warrant, but did not notify the court of this;
- The FBI knew Steele had talked with the media, but the bureau informed the court otherwise;
- The FBI cited a second source for its information, knowing it came from the original source;
- The FBI continued to "vouch" for Steele in subsequent warrants even though he had been terminated because he was no longer credible;
- FBI director Comey knew the dossier was "unverified," but signed off on the warrant applications anyway;
- The FBI knew of the political origins of the dossier, but buried this material information in a footnote in the warrant applications;
- The FBI knew that Clinton associates were also feeding Steele anti-Trump information.

These accounts uncovered by the Senate Judiciary Committee were troubling on many different levels.

The FBI's eagerness to unearth deleterious material on Trump was the driving force behind the bureau's actions to spy on one of his campaign associates, Page. In the process, the FBI and DOJ deliberately misused the FISA court to secure a warrant to spy. They deceived a succession of judges by willfully misrepresenting the information in their possession, vouching for its credibility while likely knowing full well that it was, in a word, incredible.

The Senate memo left little doubt that the FBI and DOJ tricked the FISA court by masking their knowledge that the "dossier" was nothing more than a political "hit-job." Grassley and Graham had access to the FISA warrant applications and found a stunning fact—the government was hiding something quite important:

> The FBI discussed the reliability of this unverified information provided by Mr. Steele in footnotes 8 and 18 of the FISA warrant application. First, the FBI noted to a vaguely limited extent the political origins of the dossier. In footnote 8 the FBI stated that the dossier information was compiled pursuant to the direction of a law firm who had hired an identified U.S. person—now known as Glenn Simpson of Fusion GPS—(redacted). The application failed to disclose that the identities of Mr. Simpson's ultimate clients were the Clinton campaign and the DNC.[49]

Even if a discerning judge read every single footnote, he or she would never have discovered that it was Clinton and the Democrats who had helped manufacture the sketchy "dossier" in concert with an opposition research firm, Fusion GPS. It was nowhere in print. Their identities were not disclosed.

Take a close look at footnote 8 in which the government and

Democrats insist they revealed the political nature of the "dossier" 's origins. It is a model of obfuscation:

> Source # 1 was approached by an identified U.S. Person, who indicated to Source # 1 that a U.S.-based law firm had hired the identified U.S. Person to conduct research regarding Candidate # 1's ties to Russia. (The identified U.S. Person and Source # 1 have a long-standing business relationship.) The identified U.S. person hired Source # 1 to conduct this research. The identified U.S. Person never advised Source # 1 as to the motivation behind the research into Candidate # 1's ties to Russia. The FBI speculates that the identified U.S. Person was likely looking for information that could be used to discredit Candidate # 1's campaign.[50]

A FISA judge, or any human being for that matter, would have to be telepathic to comprehend the facts underlying this well-disguised footnote: the Clinton campaign and Democrats were the ones who paid for the anti-Trump "dossier" that was manufactured by Fusion GPS and contrived by a foreign national, Christopher Steele, who based his unverified information on dubious Russian hearsay sources who may or may not exist. The FISA judges, assuming they even read the footnote, could never have discerned the truth because it was so well concealed.

The political motivations and funding behind the "dossier" were crucial and indispensable facts that should have been highlighted to the court by the FBI and the DOJ, not obscured in a couple of footnotes that never revealed a clear and honest version of how the document was generated.

This underscores just how sneaky and misleading top government officials were in their representations before the FISA judges.

The FBI and DOJ well knew that they could mask the truth and get away with it. It is a documented fact that 99 percent of all FISA applications are approved.[51] The government was aware of this and knew it would gain permission because the judges trusted officers of the court to be honest in their representations. The FISA court operates in secret. It is not an adversarial venue. No opposing views are presented because no other side is present. The integrity of the process is entirely dependent on the government being truthful and forthright.

There were other unprincipled tactics employed by the FBI, some of which would constitute outright fraud in any other circumstance or venue. Did the bureau or the Justice Department ever advise the court it had fired Steele because he "lied"? No. They prevaricated in another carefully worded footnote that he was terminated only because of an "unauthorized disclosure of information to the press."[52] This was a half-truth, at best. Did they tell the court their second source, a media story, was based on the original source? No. That inconvenient fact was omitted.[53] In short, the evidence revealed that Steele lied, then the FBI covered it up and misled the FISA court.

Senate investigators also uncovered another extraordinary fact. Not only did the Clinton campaign fund Steele's phony dossier, but Clinton's allies provided some of the bogus information contained therein:

> Mr. Steele's memorandum states that his company "received this report from (redacted) U.S. State Department," that the report was the second in a series, and that the report was information that came from a foreign sub-source who "is in touch with (redacted), a contact of (redacted), a friend of the Clintons, who passed it to (redacted)." It is troubling enough that the Clinton Campaign funded Mr. Steele's work, but

> that these Clinton associates were contemporaneously feeding Mr. Steele allegations raises additional concerns about his credibility.[54]

Clearly, the Clinton campaign and her associates were the driving force behind the dossier that was used to spy on her political opponent and investigate the Trump campaign. But Comey appears to have covered up the FBI's interactions with Steele. The Senate Judiciary Committee accused him of providing it with an account that was "inconsistent with information contained in FISA applications the Chairman and Ranking Member later reviewed."[55] The committee concluded, "it is unclear whether this was a deliberate attempt to mislead" members of Congress, and he thereafter refused "to resolve discrepancies he provided in a closed briefing and information contained in classified documents."[56]

One of the lesser known facts is that Steele himself cast doubt on the truth of his own "dossier." In a defamation case brought against him in a British court, Steele was forced to answer questions propounded by what American courts call "interrogatories," which are written questions that must be responded to under a signed "statement of truth." In criminal jeopardy if he lied, Steele admitted that his claims against Trump in the "dossier" were derived from "raw intelligence" based on "limited intelligence."[57]

Steele further acknowledged his accusations against the Trump campaign were "unverifiable" and that his sources would likely deny the allegations. In a veiled reference to the partisan motivations behind the document—the role of Fusion GPS and the Clinton campaign—he cautioned, "its contents must be critically viewed in light of the purpose for and circumstances in which the information was collected."[58] This was a subtle way of confessing that Steele's work was not worth the paper on which it was written.

In plain language, it was an untruthful political attack intended to vilify an opposing candidate.

In light of the FBI's determination that Steele had "lied" to agents, Grassley and Graham requested that the Department of Justice open a criminal investigation into Steele's conduct for potential prosecution of making false and misleading statements.[59] The senators said Steele's actions precipitated a chain of events that had a serious and harmful impact when it declared, "Mr. Steele's apparent deception seems to have posed significant, material consequences on the FBI's investigative decisions and representations to the court."[60]

ABUSE OF POWER AND OTHER CRIMES

Under the FISA law, the government was required to establish probable cause not only that Carter Page was acting as a foreign agent for Russia, but that he committed some identifiable federal crime while doing so.[61] The "dossier" purported that Page was an agent working with the Kremlin to carry out a bribery scheme and other crimes, but beyond the wild assertions in the document, there was no other independent or supporting evidence. It appears that the FBI and DOJ relied entirely on Steele and his "dossier" in their probable cause proof of both agent and crime.

Since the document itself was completely unverified, based on multiple hearsay, funded by an opposing political campaign, and compiled by a foreign national who admitted he was "desperate that Donald Trump not get elected," it was woefully inadequate for the purpose of gaining a warrant to surveil. It might have been enough for a tabloid rag at the checkout counter of a supermarket. But it came nowhere near being sufficient in any court of law in America, especially a Federal Intelligence Surveillance Court deciding whether there was probable cause to grant legal approval to engage in one of the most serious intrusions on a constitutional right—that is, to spy on a U.S. citizen.

Had the judges who signed the succession of four warrants to surveil Page known the truth about Steele and his "dossier," it is highly doubtful that they would ever have signed off on it. The judges were relying on the honesty and integrity of the FBI and DOJ officials to be candid and forthright in their representations. After all, the lawyers who submitted and argued the matter were considered "officers of the court." They were duty-bound by law and ethics to be truthful. It seems they were not. Instead, they appear to have concealed vital evidence and deceived the court.

Comey, in particular, deserves recrimination. His signature is on the early warrant applications although he admitted to Congress that he believed the "dossier" was "salacious and unverified." Within ten days of signing the first application, Steele was terminated for breaching his agreement not to speak with the media and lying about it. Comey should have immediately notified the FISA court that the FBI's confidential informant was no longer a reliable source and ended the surveillance of Page. He did not. Instead, he allowed the spying to continue and signed two more warrant application extensions that were dependent on Steele and his "dossier."

As FBI director, Comey should never have allowed such an unsubstantiated document compiled by a biased and discredited source to be utilized to spy on an associate of a presidential candidate. The surveillance continued under Comey's direction even after Trump became president.

According to congressional investigators, FBI deputy director Andrew McCabe told Congress when he testified behind closed doors in December 2017 that the bureau had verified none of the claims in the "dossier," except that Page had given a speech in

Moscow.[62] The speech, as pointed out, was not unusual and perfectly legal. The visit was also a matter of public knowledge which Steele could easily have acquired to insert in his "dossier" for the purpose of mischaracterizing it as some clandestine plot to conspire with the Russians. McCabe, like Comey, signed off on the last FISA application knowing full well its conspicuous deficiencies.

Deputy Attorney General Sally Yates and her successor, Rod Rosenstein, also signed warrant applications affirming their belief in the contents of the "dossier," its accuracy, and the reliability of their source, Steele. Surely, they also knew the legal justification for their surveillance was deeply flawed. Yet they did it anyway.

If the dossier was the primary basis for the warrant to surveil and it was nevertheless unverified, as McCabe reportedly admitted to Congress, then the legal justification for the warrant to surveil was insufficient as a matter of law. This means that all evidence collected as a consequence of the surveillance can never be used in a prosecution of any person. In what is known as the "exclusionary rule" and a companion doctrine known as "the fruit of the poisonous tree," the U.S. Supreme Court ruled that defective warrants and improper searches which violate a person's constitutional rights under the Fourth and Fifth Amendments serve as a bar to any evidence derived therefrom.[63] These rules apply specifically, but not exclusively, to cases involving wiretaps.[64]

Whatever evidence the FBI, the Justice Department, or the special counsel extracted from Page's emails, texts, and wiretapped conversations can never be used in any prosecution against him or anyone else. This effectively forecloses a "collusion" case unless evidence was acquired through other means completely unconnected to the "dossier."

However, high-ranking officials at the FBI and DOJ could and should be prosecuted for using their positions of authority to abuse the law and the constitutional rights of American citizens. At least six different felony statutes appear to have been broken.

First, it is a crime for government officials to use their position to deprive someone of their constitutional rights. Known to some as the "abuse of power" statute (18 U.S.C. 242), its formal name is "Deprivation of Rights Under Color of Law."[65] Referring to the Department of Justice analysis of this statute found on the DOJ's website, the following is stated:

> Section 242 of Title 18 makes it a crime for a person acting under color of any law to willfully deprive a person of a right or privilege protected by the Constitution or laws of the United States. For the purpose of Section 242, acts under "color of law" include acts not only done by federal, state, or local officials within their lawful authority, but also acts done beyond the bounds of that official's lawful authority, if the acts are done while the official is purporting to or pretending to act in the performance of his/her official duties.[66]

The FBI's website offers the same legal interpretation.

Some have suggested that this law applies only to cases involving discrimination against protected classes of people. This is not true. The Justice Department's published opinion makes this clear:

> It is not necessary that the crime be motivated by animus toward the race, color, religion, sex, handicap, familial status or national origin of the victim.[67]

What constitutional rights were violated? As to Carter Page, his right to privacy, affirmed by the U.S. Supreme Court as a constitutional right, was abridged by the actions of the FBI and DOJ when they spied on him. Also, his Fourth Amendment right to be secure "against unreasonable searches and seizures" in the absence of "probable cause" was violated.

Donald Trump's constitutional right to run for the public office of president may also have been violated, since it appears that the real purpose of the FBI's criminal investigation and surveillance of his former campaign advisor was to discover incriminating evidence against Trump to impede or disrupt his candidacy and, later, his presidency with the taint of Russian "collusion."

Upon conviction, "the offense is punishable by a range of imprisonment up to a life term . . . depending upon the circumstances of the crime, and the resulting injury, if any."[68]

A second potential crime is perjury as set forth in 18 U.S.C. 1621 and 1623.[69] These two statutes criminalize lying under oath (or in any statement under penalty of perjury) in federal court if a person "willfully subscribes as true any material matter which he does not believe to be true."[70]

Concealing important information regarding the "dossier" from the FISA court, such as the political source of funding, the deceit of the document's author, and the unverified nature of its contents would be obvious "material" matters. It appears to have been done "willfully" because the FBI and the DOJ were aware of this information but chose not to disclose it.

Because FISA court proceedings are held in secret, and the warrant applications are not made public, it is unclear what oaths or affirmations were made in court. If it was similar to most court proceedings, the perjury standards applied. The penalty upon conviction is a fine or imprisonment up to five years, or both.

The third criminal statute that may have been violated is often treated as an alternative to perjury called "false statements" (18 U.S.C. 1001).[71] It makes it a crime to falsify or conceal a material fact or "make any materially false, fictitious, or fraudulent statement or representation." The law also prohibits the same conduct in "any writing or document."[72] This is a broader statute than perjury because the false statement need not be under oath.

Although this law explicitly excludes judicial proceedings, it does apply to actions taken in executive branch matters. Since both the FBI and DOJ are part of the executive branch of government, any false statements made by those officials or an effort to conceal essential facts in the course of their official duties could be prosecuted as a crime.

A fourth statute that is relevant is obstruction of justice (18 U.S.C. 1503).[73] It is a crime to corruptly "endeavor to influence" a judge in any judicial proceeding. The term "corruptly" is defined as "acting with an improper purpose, personally or by influencing another, including making a false or misleading statement, or withholding, concealing, altering, or destroying a document or other information" (18 U.S.C. 1515(b)).[74] Thus, concealing from the FISA judge the political motives, money, and bias underlying the dossier could constitute obstruction of justice.

The punishment for committing this crime is incarceration up to ten years.

The fifth statute that may have been violated is known as "major fraud" (18 U.S.C. 1031).[75] This law is most often used to prosecute financial crimes, but the language encompasses a wide range of prohibited behavior. For instance, it is a crime to "knowingly execute, or attempt to execute, any scheme or artifice with the intent to defraud the United States."[76] If the FBI and the DOJ intentionally misled the FISA judge with a "dossier" they knew to be either false or unsubstantiated, this would constitute fraud, punishable by up to ten years behind bars.

A sixth statute is a conspiracy law that criminalizes an agreement to defraud the government (18 U.S.C. 371).[77] It requires proof that "two or more persons conspired either to commit any offense against the United States, or to defraud the United States."[78] A conspiracy prosecution could also be brought against the officials who worked in concert to commit any or all the above-referenced underlying offenses. The penalty reaches up to five years in prison.

These are by no means all of the crimes that may have been committed by FBI officials and lawyers at the Department of Justice if, as it appears, they intentionally misled the FISA court judges in their four requests to surveil a member of the Trump campaign based on a document they must have known was deeply flawed.

As for Christopher Steele's potential criminal liability, lying to or misleading the FBI is a felony under 18 U.S.C. 1001.[79] The Senate Judiciary Committee sent a criminal referral to the Justice Department, citing false and misleading statements.

Glenn Simpson of Fusion GPS could also be prosecuted under this same statute if it is determined that his testimony before various congressional committees was less than forthcoming, especially his apparent failure to fully disclose the role Nellie Ohr played in developing the anti-Trump opposition research.

In addition, Senator Grassley, chairman of the Judiciary Committee, sent a letter to the Department of Justice expressing his concerns that Simpson and his company, who had not registered as foreign agents, may have violated the Foreign Agents Registration Act (FARA)[80] in its pro-Russia lobbying efforts in another case:

> The issue is of particular concern to the committee given that when Fusion GPS reportedly was acting as an unregistered agent of Russian interests, it appears to have been si-

> multaneously overseeing the creation of the unsubstantiated dossier of allegations of a conspiracy between the Trump campaign and the Russians.[81]

Fusion GPS claims the money it received came from an American law firm. However, a strict reading of FARA would not protect Fusion GPS from the legal requirement that it register as a foreign agent regardless of whether compensation was delivered through an intermediary.

It cannot be overlooked that the actions of Hillary Clinton and her campaign, together with the Democratic National Committee (DNC), appear to have violated at least two laws.

First, paying a foreign national for the anti-Trump "dossier" is illegal under 52 U.S.C. 30121.[82] As explained in an earlier chapter, exchanging or donating information alone is not regarded by the law as a "thing of value" under the statute. However, *paying* for it, as the Clinton campaign did, clearly established its value, making it an illegal contribution from foreign nationals—Steele and the Russians who allegedly provided the damaging information. Fusion GPS was paid more than $1 million and Steele's company, Orbis, received $168,000 for his work in compiling the "dossier." This remuneration for a "thing of value" violated the law.

Second, not properly disclosing the "dossier" payments in reports is also a violation of the Federal Election Campaign Act under 52 U.S.C. 30101.[83] A formal complaint was filed before the Federal Election Commission in October 2017 alleging that the Clinton campaign and the DNC "failed to accurately disclose the purpose and recipient of payments for the dossier of research alleging connections between then-candidate Donald Trump and Russia, effectively hiding these payments from public scrutiny, contrary to the requirements of federal law."[84]

Many of the cases involving a breach of campaign finance laws result in civil penalties, such as fines. But in instances of *egregious* violations, criminal prosecutions and convictions have been brought by the Justice Department which has concurrent jurisdiction with the Federal Election Commission.[85] Was the conduct of Clinton and the Democrats willful and egregious? Absolutely. Paying foreign nationals for an unverified report to discredit a political opponent in a presidential campaign is about as egregious as one can imagine.

CHAPTER 8

MEETING WITH RUSSIANS IS NOT A CRIME

A Constitution is not the act of a government, but of a people constituting a government; and government without a constitution, is power without a right.

—Thomas Paine, *Rights of Man*, part 2, chapter 4

Why are foreigners on American soil allowed to volunteer their services and provide information to political campaigns in U.S. elections? The reason should be obvious—the Constitution.

The First Amendment gives Americans the freedom to associate with whomever they want, including Russians. It gives people the freedom to exchange ideas and knowledge, even with foreign nationals. As long as the information conveyed is neither stolen nor classified, there is no crime. There is no civil wrong.

We do not criminalize free speech and free association in America. Yet, there is this false mentality that all Russians are a threat to national security. Speaking to them is somehow a crime. It is not. The founders of our Constitution would be mortified at that notion.

Some have asserted that the president's son, Donald Trump Jr., had an absolute duty to notify the FBI of a meeting he had with a Russian lawyer at Trump Tower during the presidential campaign. Why? Where is that law or duty written? If it is not a crime to converse with, or obtain information from, a Russian, why should the FBI be notified? Why would agents be interested in *legal* activity? Even if a citizen learns of a crime, in most jurisdictions there is no affirmative duty imposed by law to report it to law enforcement.

As much as the media, Democrats, and some uninformed lawyers may wish it to be, it is not a crime to meet with a Russian. Nor is it a crime to meet with a Russian lawyer or government official. Even gathering information from a foreign source in a campaign is permissible. Unwise and ill-advised, perhaps. Illegal, no.

Many in the media are unschooled in the law and often oblivious to the Constitution. They became complicit in misinforming the public about the Trump Tower meeting. Rarely during an interview did a reporter or anchor ask the tough, challenging question, "What laws, specifically, were broken? Can you identify a precise statute?" The answer would have been a deafening silence.

Our Constitution was never intended to embody a political point of view. It was made for people of fundamentally differing views. Unfortunately, those in Washington who should know that do not. And the media, which owes its existence to that esteemed document, seem to have forgotten its precious guarantees of freedom.

THE TRUMP TOWER MEETING

On June 9, 2016, Donald Trump Jr. and several others in the campaign met with a Russian lawyer, Natalia Veselnitskaya, who came to Trump Tower under the pretense of having negative information about Clinton. Little, if any, discussion about the Democratic nominee occurred during the meeting. The lawyer seemed mostly interested in pressing her case that Russian adoptions in the U.S. should resume.

A year later, the *New York Times* published a story about the meeting, alleging the lawyer had connections to the Kremlin. Democrats and the media immediately seized upon it as evidence of "collusion" with the Russians and all manner of illegality.[1]

Virginia senator Tim Kaine, Clinton's former running mate, branded it "potentially treason."[2] Harvard law professor Laurence Tribe tweeted that "attempted theft of a presidential election in collusion with Putin is a serious felony and a high crime against the state."[3] Richard Painter, a former White House ethics lawyer for President George W. Bush, tweeted "this is treason" and "in the Bush administration we would have had him in custody for questioning by now."[4] Bear in mind that both Tribe and Painter sued Trump within days of his taking office claiming his business holdings violated the "emoluments" clause of the Constitution. That case was later dismissed. But it would suggest they already had a discernable bias against the new president that shaded their inflammatory condemnations.

All three of these men are lawyers who earned their degrees at Harvard or Yale. Yet, their claims have little support in the law. This underscores the observation of Erasmus, the noted classical scholar who famously dismissed lawyers as, "A most learned species of profoundly ignorant men."

Let's examine the charge. Treason is defined in Article III of the Constitution and codified in 18 USC 2381: "Treason against the United States shall consist only in levying War against them, or in adhering to their Enemies, giving them Aid and Comfort."[5]

Meeting with a Russian lawyer and, for the sake of argument, attempting to gather information, was not treason by any stretch of the imagination. Trump Jr. and other campaign associates were not levying war against the U.S., and they were not adhering to enemies of America. The U.S. was not, and is not, at war with Russia. Even if the president's son received information from the Russian government or otherwise collaborated with foreign officials, it constituted neither waging war nor aiding the enemy as the law defines it.

This was explained by Harvard Law professor Alan Dershowitz,[6] as well as Carlton F. W. Larson, a professor of law at the University of California, Davis, who is authoring a book about treason. Larson wrote the following:

> Enemies are defined very precisely under American treason law. An enemy is a nation or an organization with which the United States is in a declared or open war. Nations with whom we are formally at peace, such as Russia, are not enemies. Russia is a strategic adversary whose interests are frequently at odds with those of the U.S., but for purposes of treason law it is no different than Canada or France or even the American Red Cross.[7]

Before perpetuating the treason canard, Kaine, Tribe, and Painter should have reread the famous 1953 case of Julius and Ethel Rosenberg. They were convicted of espionage for delivering nuclear secrets to the Soviet Union. They were *not* charged with treason because the U.S. was not at war with the Soviets. Larson extrapolated an even more dire scenario that would still not constitute treason: "Indeed, Trump could give the U.S. nuclear codes to Vladimir Putin . . . and it would not be treason, as a legal matter."[8]

If meeting with a Russian lawyer or any other person connected or associated, however directly or indirectly, with the Kremlin was a crime, then a multitude of U.S. politicians who have met with Ambassador Kislyak might somehow be guilty of breaking the law. In these meetings, information is surely exchanged. No one has ever suggested that it rises to the level of treason or any other crime. Indeed, it is what diplomats and foreign officials do. It is what our officials do in foreign lands. Political campaigns are no different.

Even if the Trump campaign had acted on information allegedly offered by the Russian lawyer, it would still not constitute treason. Conspiring to subvert the government does not rise to the level of treason. Under our Constitution, Americans are permitted to voice opposition to the government, rage against political opponents, support pernicious policies, or even place the interests of another nation ahead of those of the U.S.

The only conceivable crime that might have some distant application can be found at 18 U.S.C. 371, entitled "Conspiracy to Defraud the United States."[9] In two cases, the U.S. Supreme Court broadened the statute to makes it a felony for two or more persons to enter into an agreement to interfere or obstruct a lawful function of the government.[10] An election would be a lawful government function. However, the agreement must be done by "deceit, craft or trickery, or at least by means that are dishonest."[11]

So, let's suppose that the Russian lawyer provided information that, as the email to Trump Jr. stated, "would incriminate Clinton and her dealings with Russia."[12] The Trump campaign then decided to act on the material by disseminating it to the public which would surely have a right to know. How is that deceitful or dishonest? It is not. In fact, it would be information vital to the public interest.

In any case, a campaign is allowed to utilize any information available to it, provided it is truthful, regardless of its source. Even untruthful information would not be criminally actionable, although a civil lawsuit for defamation might ensue. The Trump campaign could have repeated a truthful claim against Clinton published in Pravda, run by the Communist Party of the Russian Federation, and it would violate no law.

But this is not what happened. According to both Trump Jr. and the Russian lawyer, she offered no information about Clinton at all. In fact, the lawyer insisted the subject of the campaign was never broached. In an interview with Catherine Herridge of Fox News, Veselnitskaya said, "My meeting was not tied at all with Hillary Clinton or anyone involved with any Democrats and not at all with the presidency or the election."[13]

As for why an email sent by a business associate to Trump Jr. suggested that negative information about Clinton would be provided during the meeting, Veselnitskaya said, "I think he just exaggerated. I don't know how he came up with this idea, but he decided to use this way to get Donald Trump Jr. interested in meeting with me."[14] The email was a classic bait-and-switch—dangle a tempting pretext to insinuate a lawyer into a meeting that was really about something else entirely. It had no relevance to the campaign or Clinton.

Let's play another "what if," imagining Veselnitskaya is not telling the truth. What if she handed Trump Jr. a file and said, "Here is information that we hacked from the DNC and the Clinton campaign"? If the president's son accepted the file with the full understanding that it was illegally obtained, then it is possible he could be accused of knowingly receiving stolen property. But there is no evidence this ever happened. Moreover, the government has always been reluctant to pursue cases involving the use or publication of stolen documents or information. The government, for example, tried unsuccessfully to stop the publication of the "Pentagon Papers," but did not pursue criminal charges against either the *Washington Post* or the *New York Times* for making them public.

There is one final law to be considered. Days after the media reported on the Trump Tower meeting, Nancy Pelosi held a news conference surrounded by like-minded Democrats, during which she declared that the president's son had broken campaign laws:

> This is a campaign violation: soliciting, coordinating, or accepting something of value—opposition research, documents, and information—from a foreign national. Plain and simple.[15]

Pelosi was wrong. So were others who joined the chorus of condemnation based on laws they have surely never read.

The Federal Election Commission is the independent regulatory agency which enforces campaign laws enacted by Congress. On its government website, the FEC makes it clear that it is perfectly lawful for foreign nationals to be involved in American political campaigns:

> Even though a foreign national cannot make campaign contributions, he or she can serve as an uncompensated volunteer for a campaign or political party.[16]

The commission goes on to explain that foreigners are "allowed to attend campaign strategy meetings and events."[17] They are allowed to contribute ideas, information, and advice. They are permitted to open their mouths and speak.

None of this is considered to be a donation or "other thing of value" under the campaign statutes, as some Democrats, like Pelosi, and many in the media alleged. It is true that the Federal Election Campaign Act prohibits foreign nationals from making a "donation of money or *other thing of value*."[18] Is information, by itself, a "thing of value"? Sure, one can attach value to anything; just about everything has some measure of value. Words have meaning and, depending on the content, they often have value. But the campaign laws have never interpreted the giving of information as having value in this way. In fact, a close reading of the statute identifies specific examples of what "other thing of value" is supposed to mean in a campaign context, such as contributing to the cost of office space or electioneering.[19] Conveying information is not a prohibited contribution.

To the contrary, the Commission specifically states, "an individual may volunteer his or her personal services to a campaign without making a campaign contribution."[20]

The same language is found in both the Federal Election Campaign Act and the Code of Federal Regulations:

> The value of services provided without compensation by any individual who volunteers on behalf of a candidate or political committee is not a contribution.[21]

Thus, information is a service and is permitted to be given by a foreign national who volunteers. Jonathan Turley, a law professor at George Washington University, reinforced this point when he wrote that if information constituted a thing of value, "the wide array of meetings by politicians and their aides with foreign nationals would suddenly become possible criminal violations."[22] Constitutional lawyer Robert Barnes took it a step further when he penned a column entitled, "If Trump Jr. Is Guilty, So Is Every Democrat

Who Takes Information From 'Dreamers.'"[23] His headline and the content of his column make a valid point.

Finally, prosecutors could never bring a criminal case against Trump Jr. because they would have to show that he *knew* he was somehow breaking the law in collecting information from a foreign national. How many people know that it might be, arguably, a violation of the law? I dare say, very few. The campaign laws require what is called specific intent. That is, the person who accepts the donation must "knowingly and willfully commit a violation" of the campaign law.[24] For this reason alone, the Trump Tower meeting would not be the subject of criminal prosecution.

THE KISLYAK MEETING AT TRUMP TOWER

In May 2017, the *Washington Post* ignited a media firestorm by publishing a story that Jared Kushner and Michael Flynn met with the Russian ambassador, Sergei Kislyak, in December 2016 after Trump won the presidential election but before he was sworn into office on inauguration day.[25] The article stated that Kushner broached a discussion about "the possibility of setting up a secret and secure communications channel between the Trump transition team and the Kremlin."[26] Both Kushner and Flynn were key members of the transition team preparing for the many challenges facing the incoming administration.

Within an hour of the report, television anchors and pundits declared it a "bombshell"—their favorite description of anything related to Trump. Very few bothered to point out that nearly every recent president had established and relied on similar back-channel contacts with Moscow and other nations.

Notably, President John Kennedy depended on two sets of

back-channel communications with the Soviets to defuse the Cuban Missile Crisis in October of 1962. His brother Robert Kennedy, the attorney general, arranged an urgent deal with Soviet ambassador Anatoly Dobrynin to remove the missiles in Cuba in exchange for the U.S. removing obsolete missiles in Turkey. At the time, the State Department commandeered ABC correspondent John Scali to work out other details with Soviet Embassy official Alexander Fomin. A catastrophic nuclear exchange was averted.[27]

It made no difference whether the idea of a private communications channel was broached before or after President Trump took office. It is a distinction without a difference. As then homeland security secretary John Kelly observed, "It's both normal, in my opinion, and acceptable."[28] Richard Moss, a professor at the U.S. Naval War College, pointed out, "There's a long tradition of it—it goes back as long as diplomacy itself."[29] He was right. Back-channels were utilized in the Roosevelt, Nixon and even the Obama administrations. Nevertheless, stories in the press persisted that laws must have somehow been broken.

What Kushner did was neither improper nor different from what other incoming presidential administrations have done. It certainly did not constitute a crime.

After the election and before Trump took office, Kushner received "over one hundred contacts with people from more than twenty countries" while he was acting on behalf of the president-elect.[30] Previous entering presidents and their transition teams have engaged in similar contacts and conversations.

According to Kushner's public statement, he met with Kislyak to discuss improved relations and "to address U.S. policy in Syria."[31] Later, Kushner met with a Russian banker, Sergey Gorkov, "who could give insight into how President Putin was viewing the new administration and best ways to work together."[32] How exactly were those meetings counterproductive, uncommon, or illegal? They were not.

Professor W. David Clinton, chairman of the Political Science Department at Baylor University, who co-authored the seminal book entitled *Presidential Transitions and American Foreign Policy*, told me that he read Kushner's statement and found nothing wrong with his meetings with Kislyak.[33] Professor Clinton pointed out that it is the very same practice employed by other presidential transitions:

> It is common for representatives of other governments to get in touch with the incoming presidential administration to begin informal relationships and address relevant issues. It is not unusual. Transitions are fairly long. The incoming administration needs to inform itself of foreign policy. Getting to know people and foreign governments is widely done and beneficial to the United States.[34]

Innumerable accusations that Kushner violated the Logan Act were naive and without merit.[35] The Act prohibits private citizens from interfering in diplomatic disputes with foreign governments. As Kushner's statement explained, he did nothing of the sort. Because the meeting occurred after Trump was elected, but before he took office, the media and Democrats continued to claim that the Act was breached.

No one has ever been convicted under the Logan Act, passed in 1799, largely because lawyers, judges, and constitutional scholars regard it as an unconstitutional affront to the First Amendment.

A persuasive argument could be made that the Logan Act is no longer operable because it has remained dormant, if not dead, for more than two centuries. Prosecutors have long been schooled not to resurrect a statute that has fallowed for such a long period.

Instead, it should be allowed to rest on the books of our criminal codes as a relic collecting dust.

But even if it were somehow germane and valid, Kushner was not acting as a private citizen as the Act requires. He was acting in a wholly different capacity—as an incoming government representative of the person about to assume the presidency and the reins of the executive branch and who would be constitutionally authorized to conduct foreign affairs. Accepted historical precedent is also relevant. As Professor Clinton explained, other presidents have had detailed discussions with foreign governments before officially taking office.

Kushner was preparing the forthcoming administration for the foreign policy challenges that lay ahead and establishing the kind of vital contact that would assist the new president in formulating effective relationships and policies. In other words, Kushner was doing his job. He did it in the same manner that other transition officials have done in previous administrations.

In his appearance before congressional committees, Kushner explained that Trump had asked him to serve as the "point of contact" with foreign officials who were reaching out to discuss a variety of issues. Ambassador Kislyak wanted to discuss the fight against terror in Syria:

> He wanted to convey information from what he called his generals. He said he wanted to provide information that would help inform the new administration. He said the generals could not easily come to the U.S. to convey this information and he asked if there was a secure line in the transition office to conduct a conversation. General Flynn or I explained that there were no such lines.

> I asked if they had an existing communications channel at his embassy we could use where they would be comfortable transmitting the information they wanted to relay to General Flynn. The Ambassador said that would not be possible and so we all agreed that we would receive this information after the Inauguration. Nothing else occurred. I did not suggest a "secret back channel." I did not suggest an on-going secret form of communication for then or for when the administration took office.[36]

By his statement, Kushner disabused the media reports that he had sought some "secret back channel" to the Russian government. As it turned out, Russian generals apparently wanted to provide sensitive information on Syria that might prove helpful and sought a protected method to convey it. It was an attempt at cooperation and future coordination as the U.S. and Russia pursued a common goal—to destroy the ISIS caliphate in Syria.

During the meeting, the Russian ambassador asked if there was a secure line in the transition office to conduct such a conversation. There was not. Kushner, interested in defusing the humanitarian crisis in Syria, asked if the existing communications at the embassy could be used. It could not. This impasse ended the discussion.

Secure lines are not an extraordinary means of communicating with foreign governments. Indeed, they are conventional and unremarkable, especially when prospective military information is being shared.

Democrats might feel the Trump team was adopting a dangerous and incorrect policy. They might believe the new administration was choosing the wrong allies. That's a standard set of concerns one party has about the other. What an incoming administration does is the kind of thing elections decide. In this case, it was neither nefarious nor criminal.

The media was undeterred. It changed tactics to speculate that Kushner committed a crime by omitting his Russian meetings when he filled out his security clearance forms. But the press scarcely mentioned that people frequently commit errors and omissions in filling out the complicated security forms and are rarely prosecuted because it is exceedingly difficult to demonstrate that information mistakenly omitted was, in fact, "knowingly falsified or concealed," as the law demands.[37] Since violation is not a strict liability crime, any prosecutor would have to prove "specific intent." That is, Kushner tried to deliberately deceive the government. Incomplete paperwork, by itself, is not a crime.

Kushner explained the error in his statement to Congress. His assistant mistakenly sent on January 18, 2017, what was only an uncompleted draft. It omitted all foreign contacts, not just the Russian meetings. The mistake was noticed immediately that same day, and the transition team was notified that night. The team promptly informed the FBI the next day and thereafter delivered supplemental information that disclosed "numerous contacts with foreign officials."[38]

Still later, a complete and accurate list was given to the bureau that disclosed all foreign contacts. These events were in no way a violation of the law nor criminal. In fact, the Security Clearance form instructs the applicant that the document can be updated and clarified as part of the clearance process.[39] There was no evidence whatsoever that Kushner tried to deliberately deceive the government.

For months, Kushner was mugged by the media mob with wild and baseless accusations of criminality. In truth, his crime was no crime at all. Sadly, very few reporters made an effort to place his meetings in the context of history and precedence, as Professor

Clinton did. There was no attempt by the media at fairness or reasoned analysis.

The late U.S. Supreme Court Justice Tom Clark once wrote, "This nation was dedicated to freedom under law, not under mobs."[40] Justice Clark cared deeply about the role of the news media in holding our government accountable. But he would be dispirited to see their embrace of "mobocracy," as he once described it.

Today's mainstream media is the mob as a ruling class. They assert political control by denigrating and vilifying. No act by Trump or anyone in his administration, however slight, will be spared a full-throated "scandal" as declared by the media. All deeds are to be treated as potential crimes or impeachable offenses.

Frankly, the media deserves a good mugging. Their treatment of Jared Kushner proves it.

SESSIONS'S INTERACTIONS WITH KISLYAK

Other than a brief encounter, including a handshake, with Ambassador Kislyak at a speech in Cleveland during the Republican convention in July 2016, Senator Jeff Sessions, who campaigned for Trump, had one meeting of substance with Kislyak in his Senate office, in September 2016. The conference was requested by the ambassador because Sessions sat on the Armed Services Committee which dealt with military appropriations overseas in matters involving Russia.

Later, during his confirmation hearing to be attorney general, Sessions did not mention the meeting when asked about any contacts between Trump *campaign surrogates* and any Russians. When news broke that Sessions had met with Kislyak, House Democratic leader Nancy Pelosi held a news conference to denounce Sessions as a liar, all but pronounced him guilty of perjury, and demanded his resignation.[41] The media seized upon the meeting as evidence of Trump-Russia "collusion."

What did Sessions do to invite such condemnation? By his account, he told the truth. When asked by Senator Al Franken during his hearing about "a continuing exchange of information during the campaign between Trump surrogates and intermediaries for the Russian government," Sessions responded as follows:

> I am not aware of any of those activities. I have been called a surrogate at a time or two in that campaign and I did not have communications with Russians, and I'm unable to comment on it.[42]

In a subsequent questionnaire issued by Senator Patrick Leahy, Sessions answered "no" when asked the following:

> Several of the President-elect's nominees or senior advisers have Russian ties. Have you been in contact with anyone connected to any part of the Russian government about the 2016 election, either before or after election day?[43]
>
> Sessions's answer was "no."

As posed, the questions specifically asked Sessions about any exchange of information with the Russians by campaign surrogates and any discussion of the election itself. Following the hearing, the attorney general insisted that he responded honestly. That is, he never spoke with a Russian official on behalf of the campaign or about the election. In a statement, he said, "I never met with any Russian officials to discuss issues of the campaign. I have no idea what this allegation is about. It is false."[44]

According to Sessions, he had more than twenty-five conversations with foreign ambassadors in 2016 as a senior member of the Senate Armed Services Committee, including diplomats from Poland, Latvia, Lithuania, Hungary, Japan, Canada, and Australia.[45] Twice he spoke with the Russian ambassador, Kislyak. On a third occasion, Kislyak was present at an event in Washington, but Sessions did not recall speaking with him.

One meeting at a Heritage Foundation conference, co-sponsored by the U.S. Department of State, entitled "Global Partners in diplomacy," was held in Cleveland during the GOP convention. Foreign ambassadors to the U.S. were invited to converse with American officials, including Sessions, about national security issues, trade, and other foreign policy matters. Sessions insisted that he spoke with Kislyak only in his capacity as a senator and never about the campaign or election.[46]

The second meeting happened two months later in his Senate office. Again, Sessions maintained their conversation had nothing to do with electoral politics and everything to do with carrying out the functions of his job as a U.S. senator serving on the Armed Services Committee. Two of his senior staff members, both retired Army colonels, attended. Specific topics included combating terrorism and Russia's interests in Ukraine.[47]

At a news conference in March 2017, Sessions bristled at the notion that he had done anything wrong or had deceived Congress during his confirmation hearing:

> I never had meetings with Russian operatives or Russian intermediaries about the Trump campaign. And the idea that I was part of a quote, continuing exchange of information during the campaign between Trump surrogates and intermediaries for the Russian government is totally false.[48]

Speaking with the Russian ambassador to the United States is neither exceptional nor incriminating. After all, Kislyak is stationed in Washington. His job is to communicate with U.S. officials and to advance his country's agenda. Senators take meetings with foreign diplomats all the time.

Case in point: Democratic Senator Claire McCaskill. After criticizing Sessions for meeting with the Russian ambassador and claiming she had never done so in her ten years on the Armed Services Committee, it turned out that her tweets betrayed her. Twitter account records showed that she bragged on social media about two of her meetings with Kislyak, in 2013 and 2015.[49]

McCaskill wasn't the only one. In fact, seven then-Democratic senators convened with the Russian ambassador in a single meeting in 2013.[50] Kislyak was also a frequent visitor to the White House while Barack Obama was president with logs showing he visited at least twenty-two times.[51] Photos were unearthed showing Senator Chuck Schumer together with Russian President Vladimir Putin.[52]

As mentioned earlier, officials in the Hillary Clinton campaign also met with Kislyak, according to a Kremlin spokesman. However, almost no one accused Clinton of "colluding" with Russia.

Importantly, there was no evidence to suggest that Attorney General Sessions was untruthful in his statements. The only dishonesty came from Pelosi who deliberately misrepresented the known facts and defamed Sessions by branding him a liar. As a nonlawyer, she could have used a primer on perjury. Under federal law, 18 U.S.C. 1621, perjury is knowingly and willfully making a false statement under oath.[53] It does not take a great thinker to comprehend the plain meaning of the statute.

Sessions was acting in his capacity as a United States senator during his encounters with Kislyak, and he reiterated that he did not discuss the election or campaign. This meant his statements to the Judiciary Committee were truthful, not lies as Pelosi claimed. Armed with no evidence, she convicted Sessions without the benefit of a trial. Her accusations and slurs were not just baseless, given the facts, but contemptible.

It is easy for all to see now that Sessions broke no laws, breached no ethical rules, and "colluded" with no one. Nonetheless, at the time he was maligned and vilified by those who sought to impugn his honesty and integrity for nothing more than partisan politics. In June 2017, he was called before the Senate Intelligence Committee to defend himself and refute accusations that he had lied before Congress.[54]

Senator Mark Warner, Democrat of Virginia, set the ugly tone when he began by misrepresenting Sessions's testimony at his confirmation hearing in January, selectively reading only part of it and deliberately omitting the full question and full answer. It was a shameful cheap shot by Warner, driven either by malice or ignorance. Maybe both.

Sessions would have none of it. He expressed his contempt for what he called scurrilous accusations against him. He said his confirmation testimony had been misconstrued to form the basis for the persistent falsehood that he deceived Congress and committed perjury. Partisans had perpetuated the distortion, and the media had run wild with it.

At the hearing, the attorney general recited the full context of the question and his corresponding answer. He spelled out what he had already explained many times before. That is, he never met as a campaign surrogate with any Russian official:

> That was the context, and in that context my answer was a fair and correct response to the charge as I understood it. It

> did not occur to me to go further to other conversations or meetings.[55]

Naturally, he had prior meetings with foreign officials. That was part of his duties as a U.S. senator on the Armed Services Committee. As for "collusion" with the Russians? Sessions became visibly angry when he declared:

> I have never met with or had a conversation with Russians regarding interference with the election. The suggestion that I participated in collusion to hurt this country, which I have served with honor for 35 years, is an appalling and detestable lie.[56]

In his own quiet but deliberate manner, Sessions expressed disdain for the way he had been treated. Politicians and the media, animated by their unabashed scorn and visceral enmity toward Trump, had intentionally twisted Sessions's words to smear him and further fuel the Trump-Russia investigation.

CNN piled on the Sessions smear bandwagon when it published a story in late May 2017 accusing him of also "failing to disclose contacts he had with Russian officials" when he applied for security clearance.[57] The network quietly walked back the story months later when documents revealed that the FBI had advised Sessions's staff that he wasn't required to disclose foreign contacts that were related to his duties as a U.S. senator.[58]

Facts appeared to have gotten in the way of their reporting.

CHAPTER 9

FLYNN'S FIRING, SESSIONS'S RECUSAL, AND THE CANNING OF COMEY

It is never the law itself that is in the wrong; it is always some wicked interpreter of the law that has corrupted and abused it.

—Jeremy Bentham, *Fragment on Government*, Preface (1 Works 231)

The election of Donald J. Trump to the presidency was one of the most remarkable and improbable occurrences in American political history.

Not since the founding of our nation had someone with no previous experience in elected public office, the military, or government been elevated to the nation's highest office. Assembling a White House staff, cabinet, and leadership for the vast executive branch would have been an immense challenge for even a seasoned politician. Here, Trump was at a disadvantage. He therefore relied on many of the individuals who had joined his campaign and helped him achieve his unexpected victory. He did this out of loyalty and, to some extent, necessity. These were the people he knew and trusted.

President Trump made some choices that, in retrospect, proved unwise or ill-advised. The selection of retired General Michael Flynn to serve as national security adviser and Senator Jeff Sessions to be attorney general presented problems for the new president both before and immediately after he was sworn into office. While these two men may not have been well-suited for their respective positions, Trump could not have anticipated that mistakes by them would embroil his administration in controversies that would haunt his first year in office, and beyond.

Flynn was fired a mere twenty-four days into his tenure for allegedly being less than honest with Vice President Mike Pence in recounting conversations he had had with Russia's ambassador to the U.S., Sergey Kislyak, during the transition period. While Flynn's faulty memory of his discussions with Kislyak may or may not have justified his termination as the NSA, his interactions were lawful and consistent with what other presidential transition teams had done. The FBI and Department of Justice misinterpreted the law and misused their authority to try to entrap Flynn during an investigation that was utterly without factual merit or legal justification. The special counsel then pursued Flynn relentlessly, even though he had done nothing wrong.

Jeff Sessions's difficulties arose from an omission of fact, rather than the commission of any wrongful act. His testimony during his confirmation hearing was less than artful and complete. He either misunderstood a key question or, more likely, his answer was misinterpreted. As a consequence, Democrats on the Senate Judiciary Committee, together with a rabidly anti-Trump media, seized upon his perceived inconsistency and exploited it to pressure the new attorney general into an unrequired and imprudent recusal that would prove fateful to Trump. In disqualifying himself from matters related to Russia, Sessions made the mistake of relying on the advice of so-called ethics experts at the DOJ who misconstrued the law.

Then, there is the notorious James Comey, whom the new president inherited as FBI director. Comey had blatantly violated FBI regulations in his handling of the Clinton email case and was justifiably fired by Trump. Upon his departure, in an act of retribution, Comey unlawfully absconded with government documents and leaked them to the media to precipitate the appointment of his friend and mentor Robert Mueller as special counsel to investigate the president.

This became known as the Trump-Russia probe. In reality, it was a hoax. But the mistreatment of Flynn, the recusal of Sessions, and the firing of Comey were the flashpoints for specious claims that Trump not only "colluded" with Russians during the presidential election, but attempted to obstruct justice.

These series of events created a political upheaval that would threaten to jeopardize the Trump presidency. Had Sessions not recused himself, there would have been no special counsel, Flynn would not have been prosecuted, and the frenzied fantasy of "collusion" would have been exposed for what it was—an illusion.

But wicked interpreters of the law corrupted and abused it.

FLYNN COMMITTED NO CRIMES

Retired U.S. Army Lieutenant General Flynn spent more than three decades in a U.S. Army uniform, serving in command positions in both Iraq and Afghanistan.[1] Appointed in 2012 by President Obama as director of National Intelligence, he retired two years later amid reports of acrimonious clashes over policy with the Obama administration.[2] During the next two years, Flynn operated a private company providing intelligence and consulting services to businesses and foreign governments. Several clients had connections to Turkey and Russia. He registered only belatedly as a foreign agent by filing papers with the Department of Justice in March 2017, although his firm had previously registered with the Senate.[3]

Early in 2016, Flynn signed on as a foreign policy and national security adviser to the Trump campaign despite being registered as a Democrat. At the Republican convention, he delivered an impassioned speech criticizing what he described as Obama's "misguided" approach to international threats and attacked Hillary Clinton for violating the law in her mishandling of classified documents.[4]

After Trump's election victory, Obama reportedly warned the president-elect about appointing Flynn to a top national security position.[5] Nevertheless, Trump selected the retired general to be his national security adviser. If the Obama White House had some genuine misgivings about Flynn, they were not serious enough to end his security clearance, which was renewed in April 2016.[6] It appears that the outgoing president simply chaffed at anyone who dared to disagree with him publicly, as Flynn had done. His "warning" to Trump was probably driven by malice, not a concern over competence.

During the transition and with only three weeks left before the inauguration, the Obama administration announced sanctions against Russia in retaliation for Moscow's interference in the 2016 election. They included the expulsion of thirty-five Russian officials, as well as other measures. Obama signed the order on December 28, and it was made public within twenty-four hours. Instead of consulting with the incoming president, as other outgoing presidents had done, Obama acted unilaterally in what some perceived as a deliberate effort to disadvantage Trump.[7]

The *New York Times* described the sanctions as "the strongest American response yet," but also described Obama's Machiavellian maneuver this way:

> They also appeared intended to box in President-elect Trump, who will now have to decide whether to lift the sanctions on Russian intelligence agencies when he takes office next month.[8]

If Obama was doing whatever he could to make it difficult for his successor, he succeeded in provoking a reaction from the Trump team. The same day the sanctions were revealed, Flynn, who had already been named Trump's NSA, consulted with senior members of the Presidential Transition Team about what, if anything, should be communicated to Russia's ambassador about the sanctions.

Flynn was directed to ask Moscow "not to escalate the situation and only respond . . . in a reciprocal manner."[9] It was a reasonable and measured approach which did nothing to undermine Obama's actions, but sought to limit what might have been Russia's punishing response. Indeed, it worked. On December 30, President Vladimir Putin announced his country would not retaliate. Clearly, Flynn and president-elect Trump had managed to defuse what could have evolved into a more volatile situation.

It is uncertain whether Flynn knew it or not, but American intelligence agencies, still under the direction of the Obama administration, were listening in on his conversation with Kislyak and recording it. Armed with a transcript of it and without any legal justification whatsoever, the FBI decided to interview Flynn on

January 24, 2017, in Washington D.C., just four days after Trump was sworn in. The bureau and, later, the special counsel, then spent the next eleven months pursuing Flynn as a criminal suspect. It is unclear what Flynn said to Kislyak because their secretly recorded interaction has never been made public. Nor do we know what Flynn told the FBI, because that interview has not been made available, either. As of this writing, the FBI and DOJ have refused to divulge a transcript of the conversation despite a formal request by the Senate Judiciary Committee more than a year ago.[10]

However, in court documents filed against Flynn by special counsel Robert Mueller in December 2017, prosecutors claimed he "falsely stated that he did not ask Russia's Ambassador to the United States to refrain from escalating the situation in response to sanctions imposed against Russia" . . . and he "did not remember a follow-up conversation in which the Russian Ambassador stated that Russia had chosen to moderate its response to those sanctions as a result of Flynn's request."[11]

After battling the FBI and Robert Mueller's special counsel team for the better part of a year, Flynn was charged with making a materially false statement to the FBI under a federal criminal statute, 18 U.S.C. 1001, and agreed to plead guilty to a single count.[12] Although Flynn had previously insisted he did nothing wrong and broke no laws, he finally surrendered under the intense emotional strain and monetary pressures. His decision was understandable. Already, Flynn had been crushed financially trying to defend himself. His assets had dwindled, and he was forced to put his home up for sale to pay his mounting legal fees.[13] Had he maintained the fight to clear his name, his defense bills would have run well into the hundreds of thousands of dollars.[14] Having spent most of his professional career on a military salary, the retired general could ill afford it. He would have been left penniless and in debt.

There were also reports that Flynn felt Mueller, who was scrutinizing every aspect of the general's life, might take legal action against his son who had worked in his consulting firm and was also being targeted by the special counsel.[15] If true, Flynn's decision to throw in the towel seemed rational. Any loving father would sacrifice himself to protect a child. But these events serve as an example of how a ruthless prosecutor, if he so chooses, can exercise any means to coerce someone into copping a plea. Threatening to bring charges against a family member is an unconscionable act, but is used all too often by the heavy hand of prosecuting authorities. An individual of modest means is no match for the unbridled power and vast resources of government prosecutors. They can harass, intimidate, persecute, and break almost anyone, even an innocent person, and drive him to financial ruin. Brutal tactics by overzealous prosecutors are, sadly, endemic in the halls of justice.

Flynn was never charged with "colluding" with Russia, notwithstanding the near constant connotations advanced by a breathless media that seemed convinced he was a Kremlin sympathizer or mole. It is a shame that Flynn never had his day in court. He would likely have won the case against him, for several reasons.

First, under the relevant law, prosecutors must prove beyond a reasonable doubt that Flynn "knowingly and willfully" falsified a material fact.[16] If his recollection of the conversation with Kislyak is inconsistent with the transcript, it is not a crime. If Flynn interpreted his discussion differently than the FBI, it is also not a crime.

Let's suppose the word "sanctions" was never mentioned while talking with Kislyak, but there was a general discussion about expelled diplomats, which was part of the sanctions package. Content and context both matter. Also, the length of the meeting is critical. If the subject of sanctions was merely a passing reference in a larger conversation about moving forward to establish better

American-Russian relations in a new administration, then Flynn's reported insistence that he did not recall all of the discussion seems more credible, making it a legally viable defense.

A successful prosecution requires proof of what is called specific intent. That is, that Flynn made a false statement with the knowledge of its falsity, rather than because of confusion, mistake, or faulty memory. The defendant must have acted "willfully and knowingly." Hence, if Flynn believed his statement to be true when he made it to the FBI, even though it was arguably false, the case against him cannot be proven.

Second, the observations and impressions of the FBI agents who interviewed Flynn are not only relevant, but normally dispositive in deciding whether to even bring such a charge. Much depended on Flynn's state of mind during the interview. Thus, the agents who posed questions and listened to the answers became important percipient witnesses to the issue of his "intent." Did they walk away convinced that Flynn *believed* he was telling the truth and that any inaccuracies were unintentional? If so, no prosecution should have been brought.

So, what opinions did the agents take away from their interview? It was not until after Flynn was fired that information began to trickle out that those agents had, indeed, concluded that the NSA director told them the unvarnished truth of what he recalled from his conversations. Evan Perez of CNN reported that no charges were expected and added the following information gathered from anonymous law enforcement officials:

> The FBI interviewers believed Flynn was cooperative and provided truthful answers. Although Flynn didn't remember all of what he talked about, they don't believe he was intentionally misleading them, the officials say.[17]

Many people assumed Flynn would not be charged because the interviewing agents found the three-star general to be truthful. But months later, on December 1, 2017, Flynn surprisingly entered a guilty plea. That day, the Editorial Board of the *Wall Street Journal* dropped a stunning revelation:

> A congressional source also tells us that Former FBI Director James Comey told the House Intelligence Committee on March 2 that his agents had concluded that Mr. Flynn hadn't lied but had forgotten what had been discussed.[18]

Days later, Andrew McCarthy of *National Review* and a former federal prosecutor reported the same thing, with the following observation:

> Did (Mueller's prosecutors) decide they knew better than the experienced investigators who were in the room observing Flynn's demeanor as he answered their questions?
>
> I wonder whether Mueller's team informed Flynn and his counsel, prior to Flynn's guilty plea to lying to the FBI, that the interviewing agents believed he had not lied to the FBI.[19]

Byron York of the *Washington Examiner*, who also confirmed Comey's secret testimony, put it in context:

> To some Republicans, it appears the Justice Department used a never-enforced law and a convoluted theory as a pretext to question Flynn—and then, when FBI questioners came away believing Flynn had not lied to them, forged ahead with a false-statements prosecution anyway.[20]

How is it possible that the FBI and the DOJ could determine that Flynn told the truth but allow the special counsel to charge him with lying? More importantly, did they hide critical information from Flynn's lawyers that would have absolved their client of any criminal charge?*

The third reason Flynn should never have been charged is directly related to the second reason. The potential testimony of the FBI agents constituted what is known as "exculpatory evidence," which is any evidence the government possesses that is favorable to a defendant that would tend to exonerate him of guilt.

In the well-known case of *Brady v. Maryland* and its progeny, the U.S. Supreme Court held that prosecutors have a duty to disclose such exculpatory evidence even if they are not requested by the defense to do so.[21] It is the job, after all, of prosecutors to seek justice, not merely obtain convictions or guilty pleas under duress. Defendants are entitled to the constitutional protection of "due process" under the Fourteenth Amendment and to be treated fairly. Failing to turn over evidence helpful to a defendant is a severe violation of that treasured right, meriting a complete dismissal of the charge.

The FBI had to have known that the conclusions of the interviewing agents would be exculpatory. There is no indication that they ever disclosed it to the accused or his counsel.

The U.S. District Court Judge who presided over and accepted Flynn's guilty plea on December 1, 2017, was Rudolph Contreras. Mysteriously, he was recused from the case days later. No reason was stated; nor was any conflict of interest cited. It was kept hushed. More than three months later, reporting by the *Federalist*, as well

---

*Alternatively, some have speculated that the FBI had evidence of other crimes, and Flynn willingly pled down to this charge. See https://www.washingtonpost.com/news/the-fix/wp/2017/12/01/why-we-can-say-with-near-certainty-that-michael-flynn-has-flipped-sides-in-the-russia-investigation/?utm_term=.1646dbb317c3.

as Sara A. Carter, revealed the likely reason for the recusal. Investigators for the House Oversight Committee, digging through documents being scrupulously guarded by the Justice Department, discovered newly redacted text messages in which FBI counterintelligence chief Peter Strzok bragged about his close relationship to Contreras to his paramour, FBI attorney Lisa Page.[22] Strzok had played a major role in the Flynn prosecution since he was one of two FBI agents who interviewed the then-NSA director and determined that he had told the truth.

Why didn't Contreras recuse himself on his own, as ethics rules require? The answer may be found in the text messages themselves. In one exchange dated July 25, 2016, Page exclaims, "Rudy is on the FISC! Did you know that? Just appointed two months ago." Strzok replied, "I did. We talked about it before and after. I need to get together with him."[23] This was at a time when Strzok and the FBI were believed to have been compiling their unverified "dossier" to bring to a FISA court judge to gain a warrant to spy on Carter Page.

In another audacious text, Strzok states, "I'm in charge of espionage for the FBI. Any espionage FISA comes before him . . ." As the *Federalist* reported, "The pair even schemed about how to set up a cocktail or dinner party just so Contreras, Strzok, and Page could speak without arousing suspicions that they were colluding." Whether the dinner party ever took place is unclear. What *is* clear is that congressional investigators were convinced these texts were intentionally concealed from Congress and never turned over in an unredacted form.[24] The texts may also be evidence of attempted undue influence of a judge, which would constitute obstruction of justice and, potentially, other crimes.

The Flynn cased was handed off to another judge, U.S. District Court Judge Emmet G. Sullivan. He may have suspected something was amiss. Federal prosecutors and law enforcement are notorious for withholding exculpatory evidence. Sullivan had experience in the matter, having once presided over the corruption prosecution of former Alaska senator Ted Stevens, who was convicted. When it was later learned that prosecutors had been withholding evidence and creating false testimony, Judge Sullivan excoriated them and held them in contempt.[25] The indictment against Stevens was then set aside, vacating the conviction.[26]

After Flynn's guilty plea, Judge Sullivan ordered Mueller's team of prosecutors to produce "any information which is favorable to defendant and material either to defendant's guilt or punishment. This government responsibility includes producing, during plea negotiations, any exculpatory evidence in the government's possession."[27] Judge Sullivan pointed out that prosecutors had a responsibility to hand over to the defense any information that would have been helpful to him before he decided to enter his guilty plea.[28]

A fourth reason the charge against Flynn should not have been brought is that the FBI had no legal basis even to question him because his conversations with Kislyak were not a crime. It is the job of the FBI to investigate crimes or potential crimes. They must have a reasonable suspicion of criminal activity. Yet, as previously explained, the Trump Transition Team's interactions with the Russian ambassador were perfectly normal. Where was the crime? Nowhere. What reasonable suspicion existed? None.

Dialogues with foreign governments occur during every transition. Some are quite lengthy and involved, as the new administration prepares itself for the foreign policy challenges that lay ahead. It would be profoundly unusual if they did not happen. In no way did Flynn's discourse with Kislyak violate the Logan Act, which prohibits private citizens from interfering in diplomatic disputes between the U.S. and foreign governments.[29]

For more than two hundred years since the passage of the Act in 1799, no one has ever been prosecuted under it. The reason is quite simple—the law is regarded by most legal scholars as contrary to the First Amendment and, therefore, unconstitutional. Since no person has ever been convicted, no court has ever ruled directly on its constitutionality.

However, courts have commented on the Act from time to time. In 1964, the U.S. District Court for the Southern District of New York stated that it was likely unconstitutional because it is vague, overly broad, and ambiguous.[30] Numerous scholarly publications have argued that the Act violates the right to free speech under the First Amendment. So, its legal efficacy is doubtful.

Beyond that, the newly named national security adviser did nothing that would fall under the Act. He was serving as a representative of the incoming government, not as a private citizen as the Act requires. Moreover, Flynn did not interfere in a diplomatic dispute under the meaning of the Logan Act. To the contrary, he sought ways to de-escalate tensions over U.S. sanctions by asking the Russian government to limit its response "in a reciprocal manner." He helped to prevent an international confrontation that could have intensified into something far more serious. By doing this, Flynn was acting for the *benefit* of the U.S. government and in a manner not inconsistent with the Obama administration's wishes and policy. He could hardly be criticized for it, much less prosecuted.

In the Statement of Offense against Flynn, the special counsel also cited calls made by Flynn to Russia and several other countries requesting that foreign officials vote against or delay a United Nations Security Council Resolution.[31] However, this, too, would not fall under the Logan Act. The proposed resolution condemned Israel's settlements as a "flagrant violation" of international law.

This surely came as no surprise to the Russians or the Obama administration, since it conformed with the same public statement President-elect Trump issued the very day Flynn contacted the Russians about the pending vote.

Trump tweeted: *The resolution being considered at the United Nations Security Council regarding Israel should be vetoed.*[32]

There was nothing secretive or illegal about Flynn's communications with Moscow. And in the end, it did not matter. Russia ignored Flynn's request and voted in favor of the resolution which passed with fourteen votes.[33] The U.S., under the direction of the Obama administration, took no position on the resolution. By abstaining, it neither supported the resolution nor opposed it. And since the measure imposed no sanctions, it was nothing more than an idle diplomatic statement.

The FBI's motives in interviewing Flynn can be traced to Sally Yates, an Obama holdover who was acting attorney general until Jeff Sessions could be confirmed by the Senate and sworn into office. Yates would later be fired by Trump when she refused to have the Justice Department defend his initial executive order on immigration, almost certainly because she disagreed with the policy. Her antipathy toward the new president made her a darling of the anti-Trumpers, even though her act of defiance was nothing more than classic insubordination. The president, in his decision to terminate her, was well within his Article II powers under the Constitution to "take care that the laws be faithfully executed."[34]

However, while she was still acting attorney general, it was Yates who directed Comey's agents to interview Flynn.[35] As America's top law enforcement official, she must have known the Logan Act had no application, and she had, therefore, no legal justification for instructing the Bureau to even speak with Flynn. She ordered the agents to do it anyway, perhaps hoping to inveigle him in a way that would damage the new president.

If Yates had been acting in good faith, she would have notified the White House counsel that the FBI wanted to speak to Flynn about a matter that involved a potential crime. She did not do so, and neither did the FBI director Comey.[36] Had the White House been made aware of this critical information, it would either have rejected the interview outright for lack of legal cause or otherwise ensured that Flynn was represented by a lawyer during the interview. He was not. It was a seemingly unprincipled maneuver by Yates. But she was far from done.

Undeterred by the agents who reportedly concluded that Flynn had not lied, Yates nonetheless decided to visit the White House two days after the interview. According to her testimony before the Senate Judiciary Committee, she warned White House Counsel Don McGahn that Flynn could be vulnerable to blackmail by the Russians because he had been deceptive about his discussions with Kislyak.[37] This, of course, was a nonsensical scenario because FBI agents had determined that Flynn was telling the truth. In what way could the Russians use the truth as blackmail? Yates left out this tiny, but important, fact. Here is her account which stands as a textbook case of implying guilt by omission:

> We also told the White House Counsel that General Flynn had been interviewed by the FBI on February 24. Mr. McGahn asked me how he (Flynn) did and I declined to give him an answer to that.[38]

Why didn't Yates tell McGahn the truth—that FBI agents felt Flynn had given honest answers? Because that would have ruined the canard she was spinning that Moscow would somehow try to blackmail Flynn over a purported lie that was no lie at all. It is reasonable to conclude that Yates, who had been appointed by Obama, wanted the new president to fire Flynn, thereby wreaking havoc at the outset of his administration and generating a public scandal that would damage Trump. She seemed determined to make it happen, as her testimony revealed:

> Finally, we told them (White House Counsel's office) that we were giving them all of this information so that they could take action, the action that they deemed appropriate. I remember that Mr. McGahn asked me whether or not General Flynn should be fired, and I told him that that really wasn't our call, that was up to them, but that we were giving them this information so that they could take action, and that was the first meeting.[39]

Yates did not, however, give the White House the *full* information. She seemed to have cherry-picked what could be misconstrued as fatal to Flynn, while concealing a fact that would have likely exonerated the adviser. It was a clever, but unethical, deception. If Yates had been forthcoming, she would have admitted that the Logan Act was not violated and that FBI agents had believed that Flynn did not lie to them. She should have then apologized for commandeering the FBI for a legally invalid purpose and promptly resigned. Of course, she did not do this.

J. Christian Adams, a former Justice Department attorney, told Fox News he had no doubt about Yates's motivation:

> She saw it as her mission to sabotage the incoming administration. We know she did it on the travel ban. Now we know she did it to screw over General Flynn.[40]

The damage wrought by Yates accomplished the desired goal. Shortly after her visit to the White House, Flynn resigned as National Security Adviser amid mounting news accounts of impropriety.

Illicit or corrupt actions in the treatment of Flynn were compounded by Comey and the deception he begat in numerous public statements during his book tour. In television interviews, he repeatedly insisted that he never told Congress that FBI agents who interviewed Flynn did not believe he had intentionally lied.[41] Yet, days later the House Intelligence Committee made public a newly unredacted report that the FBI and DOJ had tried to block. It included quotes from a transcript of Comey's testimony:

> Director Comey testified to the Committee that "the agents . . . discerned no physical indications of deception. They didn't see any change in posture, in tone, in inflection, in eye contact. They saw nothing that indicated to them that he knew he was lying to them."[42]

According to the Senate Judiciary Committee in a letter dated May 11, 2018, Comey specifically told their committee the same thing—that agents who interviewed Flynn "saw nothing that led them to believe (he was) lying."[43] Was Comey lying when he testified before both congressional committees? Or was he lying during televised interviews to the American public in order to profit from his book?

The disinformation, false accusations, and wrongful prosecution of Flynn invite another significant legal question. Who leaked Flynn's conversations to the press? Whoever did so committed a crime.

THE LEAKERS COMMITTED A CRIME, NOT FLYNN

Someone in the federal government had access to the content of the Flynn-Kislyak conversations secretly recorded by American intelligence agencies. The tapes and their transcripts were classified documents. Despite this, someone appears to have conveyed them to *Washington Post* reporter David Ignatius, a highly regarded journalist, who published the information in an initial column on January 12, 2017:

> According to a senior U.S. government official, Flynn phoned Russian Ambassador Sergey Kislyak several times on December 29, the day the Obama administration announced the expulsion of 35 Russian officials as well as other measures in retaliation for the hacking.[44]

Several other stories followed in the next few weeks, including on the very day Yates visited the White House to talk about Flynn. The original and subsequent leaks were felonies by those who provided the information.

18 U.S.C. 798 states as follows:

> Whoever knowingly and willfully communicates, furnishes, transmits, or otherwise makes available to an unauthorized person, or publishes . . . any classified information concerning the communication intelligence activities of the United States or any foreign government . . . shall be fined under this title or imprisoned not more than ten years, or both.[45]

It matters not that the government leakers intended to reveal a deception by Flynn, assuming there was one (and the facts indicate otherwise). Any underlying motivation to serve the public good by disclosing a lie or misrepresentation is of no legal consequence under the statute.

Ignatius described the person who first told him of Flynn's recorded discussions with Kislyak as "a senior U.S. government official." That person committed the crime stated above. Was it Yates? Was it James Clapper, the director of National Intelligence? Both had access to the recorded conversations. But it was Yates who used the contents to engineer Flynn's firing at roughly the same time information was leaked to Ignatius and others. In their sworn testimony before Congress, both Yates and Clapper stated that they had no idea who leaked the information.[46]

The *Washington Post* published numerous articles as Yates and the FBI were targeting Flynn. In most of the stories, the reporters added new information that could have only come from government sources who had access to the classified recordings and other materials. In one such story, the *Post* revealed that the Flynn-Kislyak conversations were corroborated by "nine current and former officials, who were in senior positions at multiple agencies at the time of the calls, spoke on condition of anonymity to discuss intelligence matters."[47]

The reference to "multiple agencies" is especially disturbing. It indicates pervasive criminality across many of the sixteen agencies that make up the U.S. intelligence community, all of whom could have gained access to the secret recordings of Flynn and leaked them to the press. The government has immense resources at its disposal to aggressively investigate these classified leaks. Individuals who illegally disclosed the information should be prosecuted as the law demands. This is particularly true since the evidence suggests that the cavalcade of media reports implying criminality on the part of Flynn likely contributed to his forced departure as national security adviser.

SESSIONS MISTAKENLY RECUSED HIMSELF FROM RUSSIAN INVESTIGATION

Jeff Sessions was only three weeks on the job as attorney general when he held a news conference on March 2, 2017, to announce that he would recuse himself from any investigations involving Russia and the 2016 presidential campaign.

He should never have done so. Neither the facts nor the law required his recusal. It was a grievous mistake that relegated the attorney general to a meaningless bystander to events that threatened to overtake the new administration. It also angered President Trump, who felt Sessions should have stood firm in the face of unwarranted criticism instead of backing down.

Trump was correct in his assessment. He was entirely justified in his displeasure and frustration. Sessions had deceived his boss by concealing his intent to recuse himself from the federal investigation into Russia's meddling in the 2016 election and any connections between the Trump campaign and Moscow.

By his own admission during a Senate Intelligence Committee hearing in June 2017, Sessions began setting his recusal in motion within hours after being sworn in as attorney general on February 9.[48] On his first full day as AG, he stated that he immediately met with Justice Department officials to discuss stepping aside from the case and took preliminary steps to do so. Trump apparently had no idea this was going on behind closed doors at the DOJ. Here is part of Sessions' testimony:

> I basically recused myself the first day I got into the office because I never accessed files, I never learned the names of investigators, I never met with them, I never asked for any documentation.[49]

Clearly, Sessions was considering disqualifying himself well before he took the oath of office. It was not something that dawned on him the moment he raised his right hand in the Oval Office ceremony. And there stood the president, expecting his new attorney general to serve the nation fully and honestly. The next day, Sessions began his recusal, but did not make it official or public until three weeks later.

Failing to disclose such a material matter to the president was a serious betrayal. Perhaps the former senator from Alabama was so desperate for the job that he did not care that his recusal might undermine the presidency of the man who nominated him to be the nation's chief law enforcement officer. Or maybe Sessions was naïve in convincing himself that neglecting to disclose such a material matter was somehow inconsequential. It was not.

Trump was disgusted and angry when he learned of the recusal. At a news conference in the Rose Garden he vented:

> I am disappointed in the attorney general. He should not have recused himself almost immediately after he took office. And if he was going to recuse himself, he should have told me prior to taking office, and I would have quite simply picked somebody else. So, I think that's a bad thing not for the president, but for the presidency. I think it's unfair to the presidency.[50]

While it was unconventional, if not unprecedented, for a sitting president to publicly disparage his attorney general, Trump was right. He was entitled to know the truth, but Sessions actively hid it from him. He deserved an attorney general who, at the outset, was forthright about his intentions, not someone who was concealing his plan to step aside from a major investigation that would surely impact the new administration. In so doing, the attorney general "sandbagged" the president. Sessions's deception also deprived him of Trump's confidence and trust which are essential to the job of attorney general. This ethical impropriety rendered Sessions unfit to serve.

But for Sessions's deceit, it is unlikely that a special counsel would have been appointed. Instead, Sessions's replacement in the Russia case, acting attorney general Rod Rosenstein, took it upon himself to appoint Robert Mueller to preside over the probe. This played neatly into the scheme admittedly devised by fired FBI director James Comey, who just happened to be Mueller's close friend and long-time professional ally. This will be explored in the next section of this chapter.

Sessions told the Senate Intelligence Committee he was "required" to recuse himself under the Code of Federal Regulations, 28 CFR 45.2.[51] However, this was an incorrect reading of the rules. The language of this section offers broad latitude and discretion for Sessions to have remained involved. The ethical standard is a subjective one. Moreover, a strict reading of the regulations does not cover a matter like the Russia probe as it existed at the time of his recusal.

When Sessions was sworn in, the FBI's examination of Russian meddling was officially a counterintelligence case, not a criminal investigation, according to Comey's sworn testimony.[52] It is a critical distinction. Here is how the regulation reads, in relevant part:

> No employee shall participate in a criminal investigation or prosecution if he has a personal or political relationship with:
>
> 1. Any person or organization substantially involved in the conduct that is the subject of the investigation or prosecution; or

2. Any person or organization which he knows has a specific and substantial interest that would be directly affected by the outcome of the investigation or prosecution.[53]

The operative words in the regulation are "investigation or prosecution." At the moment of Sessions's recusal, neither of these two conditions were present. If the bureau's investigation had, in fact, been criminal when Sessions took office, then, arguably, he might have been correct in recusing himself. But it was not.

At roughly the same time the attorney general disqualified himself, FBI director James Comey told the House Intelligence Committee that the investigation into Russian meddling was a "counterintelligence" case which involved information gathered by law enforcement to protect against espionage.[54] Significantly, the case had not advanced to the stage of an official criminal probe, much less a prosecution, as the recusal regulation prescribes.

Former federal prosecutor Andrew McCarthy published two excellent columns explaining the distinction and then concluded as follows:

> Counterintelligence investigations should not trigger disqualification or recusal of an attorney general unless and until the investigation turns up incriminating evidence that could form the basis for a criminal investigation—and a possible prosecution.
>
> It was unnecessary for Sessions to recuse himself from the broader counterintelligence investigation. It is not a criminal inquiry, and Regulation 45.2 explicitly applies only to criminal investigation or prosecution.[55]

★ ★ ★

McCarthy's point was that Sessions relied on a recusal rule that applied to an altogether different kind of investigation, not the one taking place when the attorney general mistakenly disqualified himself. Assuming Comey's description of the case to Congress was accurate, there was no evidence of criminality connected to the Trump campaign and, thus, no need for Sessions to have removed himself, as McCarthy explained.

What about Flynn, who had been questioned by the FBI and fired by Trump? This had nothing to do with Russia's interference in the election and no correlation to activities of the Trump campaign. It was triggered by the FBI's and DOJ's misapplication of the Logan Act, as noted earlier, seemingly unconnected to a counterintelligence probe by the bureau. Sessions might have been compelled to recuse himself from the Flynn investigation, but there was no requirement that he step aside from a counterintelligence investigation related to Russia.

When Sessions consulted officials at the Justice Department the day he arrived, he may well have been misled by those Obama holdovers who had been running the department for the last eight years. Were their recommendations that he disqualify himself influenced by their political convictions? Quite possible. It is also conceivable that they did not bother to apprise themselves of the true nature of the FBI's investigation. Had they done so, they would have realized that a counterintelligence matter in no way required disqualification of the attorney general under the relevant regulation.

Sessions's contacts with Kislyak were negligible. Their only meaningful discussion during the election occurred when Sessions was acting in his official capacity as a U.S. senator, not as a surrogate of the Trump campaign. He insisted they never spoke about the election. Why then did he feel compelled to recuse himself from a matter involving Russia's interference in the election? If he answered all questions truthfully in his confirmation hearing and was, therefore, uninvolved in any alleged "collusion" with the Russian ambassador, there was no valid reason for him to disqualify himself from overseeing the FBI's counterintelligence probe. Instead, Sessions capitulated to the demands by Democrats (and some Republicans) for his removal which were based on misrepresentations of their meeting. Sessions unwittingly played neatly into the hands of Trump's political adversaries.

The attorney general recused himself from a case that, by legal definition, did not involve a crime that can be found anywhere in America's criminal codes. Indeed, there was no criminal investigation at the time he disqualified himself. Did he not read the statutory law? Did he not comprehend the recusal regulation he cited? What prosecutor removes himself from a criminal investigation that might not happen and based on a law that does not exist? Sessions reacted incorrectly to a politically driven narrative that had no support in the law.

Other attorneys general have presided over investigations involving potential crimes within their administrations. Attorney General Loretta Lynch, for example, oversaw the FBI's criminal investigation of Hillary Clinton in its entirety. Even when her involvement was revealed in the infamous tarmac meeting with Bill Clinton, she did not recuse herself. Instead, she said she would accept the recommendations of the FBI. In her testimony before the House Judiciary Committee, she admitted that she was the person who "made the decision" not to charge Clinton with any crimes.[56]

If ever there was an instance in which an attorney general should have recused herself under Regulation 45.2, it was the Clinton case. Yet Lynch declined to do so despite her private and personal contact with the husband of the subject of the investigation.

By comparison, Sessions had no known contact with anyone suspected of being involved in Russian meddling in the 2016 campaign. Yet, he recused himself in the confused or misguided belief that he was required to do so. Lynch was obligated under the law, but Sessions was not.

Sessions can be legitimately criticized for a variety of failures. He resisted producing documents involving the anti-Trump dossier which were subpoenaed by the House Intelligence Committee. He spent months ignoring pleas from members of Congress to reopen the case against Hillary Clinton for likely violations of the Espionage Act in the use of her private email server, obstruction of justice for destroying more than 30,000 emails under congressional subpoena, and potential self-dealing for profit through her foundation. His efforts to investigate the unmasking of Trump aides during intelligence surveillance by the Obama administration appeared less than vigorous. His reluctance to investigate the possible criminal conduct of James Comey is baffling, given substantial evidence that the fired FBI director may have falsely testified before Congress, stolen government documents, and leaked them to the media. Of course, contorting the law to clear Clinton was deserving of its own investigation.

Sessions may have been a decent United States senator. But in his first year in office, he proved to be an ineffectual feckless attorney general whose only real achievement seemed to be alienating the president. Amid Trump's public denunciations of Sessions, he offered a letter of resignation which was, regrettably, not accepted by the commander in chief.[57] Trump was right when he said that choosing Sessions as attorney general was one of the "worst" decisions he had made.[58]

Among Sessions's many mistakes, recusing himself from the Russia investigation had the most serious consequences for the

Trump presidency and the country. Had he maintained his authority over the matter, it is doubtful that a special counsel would ever have been appointed. And Comey's manipulations upon being fired would have been the subject of an appropriate investigation by someone other than the department's inspector general who has no jurisdiction to compel Comey and other departed FBI agents to answer questions or produce documents.

THE FIRING OF JAMES COMEY

James Comey should have been fired the day he held a very public news conference in July 2016 announcing his recommendation that Hillary Clinton not be criminally prosecuted for mishandling classified information and jeopardizing national security. He acted without authorization and in dereliction of his duty to follow the policies and regulations established by both the FBI and the Department of Justice. In so doing, he demeaned the work of the agency he led, damaged the integrity of the nation's premier law enforcement agency, and breached the public's trust.

In various congressional appearances, Comey offered a variety of excuses for why he thought extraordinary circumstances demanded him to engage in extralegal actions.[59] But none of them justified violating the fundamental rule of law as Comey did.

Within two weeks after the U.S. Senate confirmed Rod Rosenstein as deputy attorney general, he authored a report that recommended to the president that Comey be fired. The FBI director reports directly to the deputy AG. So, in that context, President Trump acted expeditiously in accepting Rosenstein's counsel.

The president appeared to think this move would be treated as a win-win. Over the previous year, both parties had accused

Comey of incompetence, grandstanding, and attempting to sway the presidential election one way or the other. This early in his tenure, President Trump still underestimated the Democrats' fecklessness.

Democrats, of course, were quick to condemn the president for discharging Comey. Many of them were the same politicians who denounced the director in late October 2016 for abusing his authority and unduly influencing the election when he sent a letter to congressional leaders briefly reopening the Clinton case only days before the election. But in Washington, hypocrisy is endemic and memories are fleeting.

However, the greatest hysteria over Comey's firing emanated from the media. Comparisons to Richard Nixon's infamous "Saturday Night Massacre" abounded, with accusations that Trump used the bungled Clinton case as an excuse for firing Comey and that the president's true purpose was to quash the Trump-Russia probe. This demonstrated that most talking heads know little about presidential history and how FBI investigations work. The director's departure did not stop the bureau's ongoing probe into Russian meddling and any alleged "collusion." It was Comey who was dismissed, not the career professionals at the agency. The counterintelligence case continued and so, too, did Congressional investigations into the same matter. They did not miss a beat. But the media frenzy advanced unabated because so many who occupy newsrooms across America harbor unabashed contempt for Trump and his unconventional approach to governing.

Comey's lack of integrity and defiance of rules and principles of law were his downfall. His unchecked ambition and desire to thrust himself into the public limelight only exacerbated his mistakes of judgment and deed.

Rosenstein's scathing critique of how Comey abused his authority, usurped the power of the attorney general, and broke FBI/DOJ rules is worth reviewing:

> I cannot defend the Director's handling of the conclusion of the investigation of Secretary Clinton's emails, and I do not understand his refusal to accept the nearly universal judgment that he was mistaken. Almost everyone agrees that the Director made serious mistakes; it is one of the few issues that unites people of diverse perspectives.
>
> My perspective on these issues is shared by former Attorneys General and Deputy Attorneys General from different eras and both political parties.[60]

In his three-page summary, Rosenstein detailed multiple acts of misconduct by Comey. The FBI director took it upon himself to assume the powers and duties of the attorney general in deciding the Clinton case and chose to announce his findings, without notice to the Justice Department, in a nationwide news conference. This was in clear contravention of regulations that restrict certain conduct by FBI employees. As Rosenstein noted, "the FBI Director is never empowered to supplant federal prosecutors and assume command of the Justice Department. The Director announced his own conclusions about the nation's most sensitive criminal investigation, without the authorization of duly appointed Justice Department leaders."[61]

Comey should have determined whether there was evidence to justify prosecution, then turned the matter over to federal prosecutors to decide whether an indictment of Clinton was warranted. Rosenstein cited the opinions of six former top Justice Department officials who all agreed. Comey's obstinate refusal to admit his errors only reinforced the need to fire him. In a stinging rebuke,

Rosenstein added, "The FBI is unlikely to regain public and congressional trust until it has a Director who understands the gravity of the mistakes and pledges never to repeat them."[62] Hence, Comey was canned for reasons that were entirely justified, an apoplectic media and howling Democrats notwithstanding.

Almost immediately, demands for impeachment of President Trump were heard in the corridors of Congress. The liberal media were crazed with excitement over the prospect that the president had obstructed justice in trying to block the Russian investigation. In truth and in law, neither scenarios were remotely rational.

CHAPTER 10

OBSTRUCTION OF JUSTICE

Men are more often bribed by their loyalties and ambitions than by money.

—Justice Robert H. Jackson,
United States v. Wunderlich, 342 U.S. 98 (1951)

There is nothing wrong with ambition. But ambition without character or principles is what turns good men into bad men. They well know right from wrong, but are too willing to hide the truth in favor of advancing their own designs. Their loyalty is to themselves alone.

It is, therefore, ironic that James Comey, the fired FBI director, penned a tome entitled *A Higher Loyalty: Truth, Lies, and Leadership*.

The *real* truth is that Comey's unchecked ambition is what led to his demise. It is the classic story of how power corrupts.

Comey took it upon himself to decide Hillary Clinton's fate in the email scandal that bedeviled her 2016 presidential campaign. He presumptuously exceeded his authority, mangled the rule of law, and arrogated partisanship over principles. His superiors at the Department of Justice eventually condemned his actions.

They found that he "flouted rules and deeply engrained traditions, and guaranteed that some people would accuse the FBI of interfering in the election."[1] Indeed, many Americans did. For this, he was appropriately fired.

But Comey was unchastened and determined to avenge the indignity of his dismissal. The notion of how to do it came to him, he said, when he awakened "in the middle of the night."[2] He resolved to leak government documents he had furtively taken from FBI headquarters to harm President Trump. It was an act of shameless hypocrisy, since Comey had promised the president, "I don't do sneaking things, I don't leak, I don't do weasel moves."[3] Thus, the man who had previously been asked to enforce the law against leakers, became a notorious leaker himself.

He also became the chief witness in a politically motivated, albeit misguided, attempt to accuse the president of obstructing justice. Comey fashioned himself the innocent victim in what became, at its core, a clever plan to incriminate Trump for words and deeds that bore no resemblance to the law of obstruction.

TRUMP DID NOT OBSTRUCT FLYNN CASE

The day after the February 13, 2017, resignation of national security adviser Michael Flynn, a national security meeting was convened in the Oval Office at the White House. Vice President Mike Pence, FBI director James Comey, Attorney General Jeff Sessions, and other senior security officials attended. According to Comey, President Trump asked the FBI director to remain behind as others departed. A conversation allegedly ensued in which the Flynn case was discussed. Here is Comey's account as told to the Senate Select Committee on Intelligence when he testified four months later:

> The President then returned to the topic of Mike Flynn, saying, "He is a good guy and has been through a lot." He repeated that Flynn hadn't done anything wrong on his calls with the Russians, but had misled the Vice President. He then said, I hope you can see your way clear to letting this go, to letting Flynn go. He is a good guy. I hope you can let this go. I replied only that "he is a good guy." (In fact, I had a positive experience dealing with Mike Flynn when he was a colleague as Director of the Defense Intelligence Agency at the beginning of my term at FBI.) I did not say I would "let this go."[4]

Comey's story of what Trump is supposed to have said was contained in a memorandum composed by the director. After he was fired, he arranged for it to be delivered secretly to the *New York Times*. Comey did not want to leak it himself, so he gave his memo to a friend—a law professor—with instructions that it be surreptitiously conveyed to the media.[5] It was a quintessential maneuver by Comey—sneaky, underhanded, and self-serving. He wanted to damage Trump after being sacked, but didn't want to dirty his own hands with the actual deed.

Later, during his Senate Intelligence Committee testimony, Comey came clean, perhaps realizing that he would eventually be found out and did not want to run the risk of perjuring himself. But along with his admission, he offered an astonishing explanation for his motivation: he did it for the sole purpose of triggering the appointment of a special counsel in a blatant effort to manipulate the legal process against Trump.[6] He bragged about what he'd managed to accomplish. If ever there was an act of retribution or vengeance, this was it. But it also constituted an apparent crime by Comey, which will be discussed later in this chapter.

Thanks to Comey's engineered leak, the mainstream media immediately ran stories claiming that Trump's remarks were evidence of obstruction of justice for attempting to derail any investigation into Flynn's conduct.[7] But wait, what investigation? When Comey met with Trump, the director already *knew* that Flynn had told the truth when he was interviewed by FBI agents in January. The investigation should have been closed. Why wasn't Comey honest and forthright with Trump by telling him that Flynn had largely been cleared by his agents of any wrongdoing? Flynn's conversations were perfectly legal and not a violation of the Logan Act, which Comey surely knew. And Flynn's responses to the agents who questioned him were deemed by them to be truthful, which Comey also knew.

Comey's unconscionable deception leaves three possibilities. First, there never was a conversation with Trump about the fired NSA, meaning Comey invented it to harm Trump. Second, Comey embellished his account of what was said. Or, third, there was a discussion about Flynn facing legal jeopardy, but Comey concealed the fact that Flynn had done nothing that violated the law. In each case, Comey appears to have been dishonest.

The White House responded to Comey's leak by denying the conversation, stating "while the president has repeatedly expressed his view that General Flynn is a decent man who served and protected our country, the President has never asked Mr. Comey or anyone else to end any investigation, including any investigation involving General Flynn." The statement went on to say, "This is not a truthful or accurate portrayal of the conversation between the President and Mr. Comey."[8] Months later, Trump repeated his denial when he tweeted, "I never asked Comey to stop investigating Flynn. Just more Fake News covering another Comey lie!"[9]

Comey's testimony about his conversation with Trump prompted widespread speculation among Democrats and the media that the former director's good friend Robert Mueller would use his appointment as special counsel to bring charges against the president for obstruction of justice or, at the very least, a report that could be used for impeachment proceedings by Congress. Senator Dianne Feinstein, who is not a lawyer but is the top Democrat on the Senate Judiciary Committee, surmised, "what we're beginning to see is the putting together of a case of obstruction of justice."[10] Feinstein should have gone to law school.

Obstruction of justice is defined in a series of statutes, the most relevant of which is 18 U.S.C. 1505:

> Whoever corruptly . . . influences, obstructs, or impedes or endeavors to influence, obstruct, or impede the due and proper administration of the law under which any pending proceeding is being had before any department or agency of the United States . . . shall be fined under this title, imprisoned not more than 5 years, or both.[11]

A "pending proceeding" would include an FBI investigation. As noted in chapter 2, the operative word in the statute is "corruptly." Because it is a somewhat fungible term that can be subject to varying interpretations, it was given an explicit meaning in a subsequent code section, 18 U.S.C. 1515(b), which also included examples of specific acts that would constitute behaving "corruptly":

> As used in section 1505, the term "corruptly" means acting with an improper purpose, personally or by influencing an-

> other, including making a false or misleading statement, or withholding, concealing, altering, or destroying a document or other information.[12]

Assuming Comey's account of what Trump said is true on its face, the president's words or actions do not satisfy any of the examples of a "corrupt" or "improper" purpose stated in the statute. No false statements, no withholding or concealing, and no altering or destroying evidence have been alleged by Comey or anyone else. Did Trump do any of those things or ask the director to do them? Reread Comey's statement. The answer is no.

Obstruction of justice is a specific intent crime, which means a person must specifically intend to obstruct or interfere with an investigation. Proof is dependent on the precise consciousness and purpose of the actor, in this case, Trump. There was nothing specific in the words the president allegedly used. His allusions to Flynn were vague and ambiguous. They could be interpreted in a variety of ways and with different meanings.

Obstruction is an easy accusation for someone like Feinstein and the media to throw around, but it is exceedingly difficult for a prosecutor to prove in a court of law because the wording of the statute is exact and unequivocal. Indeed, courts have interpreted it quite narrowly. The U.S. Supreme Court, in the case of *Arthur Andersen v. United States*, defined the term "corruptly" as it applies to a companion obstruction statute (18 U.S.C. 1512) as follows: "wrongful, immoral, depraved, or evil."[13] The high court ruled that consciousness of wrongdoing is required.

Let's compare this to the words allegedly uttered by Trump. The president is supposed to have said, "I hope you can see your way clear to letting this go, to letting Flynn go. He's a good guy. I hope you can let this go." Trump's statement bears no resemblance to the requirements of the statute or the Supreme Court explanation of a corrupt purpose. There is no lie, threat, or bribe, and no pernicious act. Trump's words were not "immoral, depraved or evil" in any sense. Under no legal interpretation could the president have obstructed justice.

Former Special Assistant U.S. Attorney at the Department of Justice Fred Tecce argues that the president's remarks do not come close to satisfying the legal requirements of obstruction of justice:

> At the time of the conversation with Comey, the FBI Director was aware that Flynn had been truthful with the FBI—Trump didn't know that, but Comey did. So, it begs the question as to how Trump could form the requisite intent to corruptly influence an investigation that at that point was not even an investigation or, at best, had revealed that Flynn had done nothing wrong.
>
> Trump's words were aspirational. How Comey took them is of no avail. The question is Trump's state of mind at the time and these words fail to indicate any intent to corruptly obstruct or impede anything.[14]

Hoping or wishing for an outcome is not the same as directing or ordering someone to end an investigation or clear a suspect.

By contrast, if the president had said, "Bury whatever incriminating evidence you might have, exonerate Flynn and end the investigation of him entirely or you are fired," that might, *arguably*, constitute obstruction of justice. It would include two "corrupt" elements—a threat and concealing evidence. It would also be regarded as an edict or mandate to stop the Flynn investigation.

Of course, this analysis ignores the fact that the president denied he ever spoke the words ascribed to him. With no known witnesses to corroborate Comey's claim, a prosecutor would be loath to consider bringing such a case based on one person's word. It is the definition of reasonable doubt.

President Trump's remarks about Flynn, if true, were not a directive to drop the investigation of him. It was not even a request, although Comey claimed he perceived it to be. His perception of what he heard is not, under the law, what proves obstruction; it is the specific intent of the speaker. In his introductory statement to the Senate Intelligence Committee, Comey asserted, "I had understood the President to be requesting that we drop any investigation of Flynn in connection with false statements about his conversations with the Russian ambassador in December."[15] How Comey interpreted the president's words is irrelevant. As the statute makes plain, the intent of the person speaking the words proves obstruction, not how the listener may construe it.

Comey was schooled in the law and knew that his story of the Oval Office meeting was not obstruction of justice at all. During his Intelligence Committee appearance, Senator James Risch posed the key question:

> **SEN. RISCH:** Do you know of any case where a person has been charged for obstruction of justice or, for that matter, any other criminal offense, where they said or thought they hoped for an outcome?
>
> **COMEY:** I don't know well enough to answer. The reason I keep saying his words is I took it as a direction.[16]

Comey's answer was a misdirection. He claimed he really couldn't answer, then shifted to how he interpreted the president's words, which is irrelevant under the obstruction statute. Senator Risch continued to pursue the issue:

SEN. RISCH: You don't know of anyone ever being charged for hoping something, is that a fair statement?
COMEY: I don't as I sit here.[17]

Intelligence Committee chairman Senator Richard Burr asked Comey quite directly whether he thought the president was trying to obstruct justice. As expected, Comey demurred by claiming, "I don't think it's for me to say whether the conversation I had with the president was an effort to obstruct."[18] His response was curious. Comey was all too willing to usurp the authority of the attorney general in deciding that Clinton broke no laws, but was reluctant to say whether Trump did. While there was no legal basis for declining to answer the pivotal question, Comey dodged the answer for a reason. If he had said, under oath, that he regarded the president's words as obstruction, Comey might have incriminated himself in a crime known as "misprision of felony" (18 U.S.C. 4).[19]

Misprision of a felony is similar to aiding and abetting a crime or accessory after the fact. It requires a showing that someone has *knowledge* of a felony and *conceals* it. What constitutes concealment? The case most often cited, *Neal v. United States,* stated: "there must be a concealment of something such as *suppression* of the evidence or other positive act."[20] In that case, evidence of a theft crime was allegedly stored or concealed in a golf bag.

Comey composed his memo and stored it somewhere until he was fired as FBI director. He either kept it at his home or took it with him when he was terminated. Either way, he considered it to be of significant value. One could argue that he regarded the memo as evidence of a felony (obstruction) and took affirmative steps to *suppress* it until he later chose to leak it to the media.

Which is more likely: Comey thought he would get away with misprision of a felony, or he thought the memo looked bad enough to be used as a political shield if he's ever fired? Why did he wait several months and only after he was cashiered to act on his memo?

People who serve in law enforcement have a special duty to immediately report knowledge of a felony to a person of higher authority, not simply bury it somewhere in a file or take it home to use sometime in the future for blackmail or retribution. The FBI director's superior is the deputy attorney general at the Department of Justice. In his testimony, Comey admitted he did not inform anyone at the DOJ.[21] His excuse was that he did not think the deputy AG would be in that position for very long, which is no excuse at all. Thus, it is reasonable to conclude that Comey did not believe the president obstructed justice and he, therefore, had no duty to report it.

Comey was pressed on this issue during his hearing. Unbelievably, he claimed he did not know whether FBI officials have a duty to report a crime that has been committed:

> **QUESTION**: You're unsure whether they would have a legal duty?
>
> **COMEY**: That's a good question. I've not thought about that before. There is a statute that prohibits misprision of a felony—knowing of a felony and taking steps to conceal it. But this is a different question. Let me be clear, I would expect any FBI agent who has information about a crime to report it.[22]

Comey expected his agents to report evidence of crimes, yet he did not do so himself. He filed away his memo, took no action, and never informed the Justice Department. His actions indicate he *knew* his memo was not evidence of a crime. He well recognized that the president's comments about Flynn did not have a "corrupt" or "improper purpose" and were never intended to obstruct or impede an investigation that had already found no wrongdoing by the national security adviser.

Comey's memo was treated as a "smoking gun" only because the media and Democrats, prompted by the former director himself, peddled it that way. At most, the Oval Office conversation was an uncomfortable encounter. But that did not stop Comey from exploiting it later to smear the man who fired him. In this respect, the ex-director succeeded.

Although the president's supposed remarks about Flynn appeared to be more of an observation than an order, Trump could have ordered Comey to end any criminal pursuit of his former NSA and been acting well within his authority to do so. This was conceded by Comey during his testimony when he endorsed what many constitutional scholars have long maintained. That is, the president has broad constitutional powers to stop investigations and potential prosecutions:

> I am not a legal scholar, but as a legal matter, the president is the head of the executive branch and could direct, in theory . . . anyone being investigated or not. I think he has the legal authority. All of us ultimately report in the executive branch to the president.[23]

On this point, Comey was right. Article II of the Constitution contains what is known as the "vesting clause." It states, "The executive Power shall be vested in a President of the United States."[24] This means he has full power over the executive branch of government and his decisions are not subject to congressional restraint. He can direct any agency or department to take whatever course of action he deems appropriate or, in the alternative, to refrain from taking action. In this case, Trump was empowered to order Comey to end his pursuit of Flynn, although that is not what he did.

Alan Dershowitz, professor at Harvard Law School and a prominent constitutional scholar, argued that Trump's actions were perfectly permissible and historically supported:

> Until recently, presidents—from Adams to Jefferson to Lincoln to Roosevelt to Kennedy—played active roles in deciding who to investigate and prosecute. In recent years a tradition had developed under which the FBI and the criminal division of the Justice Department were more independent of the White House. But this tradition did not and could not limit the constitutional authority of the president, especially in the absence of legislation.
>
> It is clear, therefore, that Trump acted within his constitutional authority if he directed Comey to end his investigation of Flynn. It follows from this that he certainly acted within his authority if he merely requested or hoped that Comey stand down.[25]

Dershowitz was wise to emphasize the importance of presidential precedent. No one ever accused President Obama of obstructing justice when, in a televised interview on Fox News, he acknowledged that Hillary Clinton was "careless" in mishandling classified documents, but insisted she did not jeopardize national security.[26] This was a thinly disguised statement to the FBI and the Justice Department that he did not want her to be prosecuted. He did not "hope" or "wish" that she be cleared. Obama articulated a very specific statement tailored to the law that his former secretary of state should be absolved of any wrongdoing. He didn't need to issue a direct order to Comey or Attorney General Loretta Lynch. His nationally televised words conveyed the unmistakable message not to bring criminal charges against Clinton, incriminating evidence notwithstanding.

Right on cue, Comey reacted to Obama's statement as the president surely intended. The director adopted Obama's argument, in nearly identical language, when he announced that Clinton had been "extremely careless" but had not "intended to violate laws governing the handling of classified information."[27] Comey acceded to the president's wishes. Yet, never once did Comey complain that Obama was interfering in, or obstructing, a potential prosecution, although the president's intent that the case be dropped can plainly be inferred from his words.

If Obama did not obstruct justice, neither did Trump. Under the First Amendment, both were entitled to express their opinions about the respective cases and the individuals involved. Indeed, under the "vesting clause," both presidents could have been far more aggressive in ordering an end to the probes. They would have been well within their constitutional powers to do so.

The only reason Trump has been accused of obstruction is because Comey, Democrats, and the liberal media shaped it into one of their favored anti-Trump narratives. It is part of a much larger narrative casting Trump's unusual style as breaking the law, when he's only breaking the establishment's unwritten rules.

In making decisions on who should or should not be prosecuted, the FBI and DOJ have enormous latitude. So, too, does the president, who can command the direction of their investigations and prosecutions or instruct that cases not be pursued. This is a legitimate executive function derived from the "take care" clause in Article II of the Constitution, which states that the president "shall take Care that the Laws be faithfully executed."[28] Such decisions invariably involve judgment and discretion. Eugene Kontorovich, law professor at Northwestern University, made this point when he observed:

> No laws are enforced 100 percent. Lots of people commit crimes; lots of prosecutors know about them and decide not to charge those crimes. Executive discretion and prioritization are part of the justice system.[29]

A prime example of such discretion can be found in President Obama's executive action to shield millions of undocumented immigrants from prosecution and deportation.[30] Was the president obstructing justice by actively interfering in potential prosecutions? No. While it is true that litigation ensued to reverse the president's order, no one argued that his action to halt prosecutions constituted obstruction under the law. Whereas Obama had issued a written order directing that prosecutions be deferred or dropped, Trump merely expressed a "hope" that Flynn would be cleared. The same mainstream media and Democrats who applauded Obama for his expressed order condemned Trump for his vague and elastic comments to Comey.

Former Assistant U.S. Attorney Doug Burns dismisses the notion that Trump obstructed justice in his comments to Comey about Flynn:

> The Trump statement to Comey was certainly ill-advised; however, it did not amount to obstruction of justice. First, the statute requires evidence of a corrupt intent, and nothing in the statement provides any such evidence. Second, the president can direct the Justice Department to drop a case as he is the attorney general's boss. Last, the president can

pardon Michael Flynn. As such his statement alone does not amount to obstruction of justice.[31]

There was no evidence the president acted "corruptly" or with an "improper purpose," as the law of obstruction requires. Those who accused him of it either misunderstood the statutes and case law or deliberately misinterpreted them for political reasons.

FIRING COMEY WAS NOT OBSTRUCTION OF JUSTICE

President Trump's firing of Comey did not constitute obstruction of justice for many of the same reasons that his conversation about Flynn did not rise to the level of obstruction. The president is constitutionally authorized to fire any political appointee in the executive branch of government. They report to him and serve at his pleasure.

The best evidence of this came from the words of the man who was fired. The day after he was terminated as FBI director, Comey wrote a letter to his colleagues at the bureau that stated:

> I have long believed that a President can fire an FBI Director for any reason, or for no reason at all.[32]

Comey was correct. Under the Constitution, the president has unfettered power to discharge the FBI director or anyone else in the executive branch who was politically appointed to his or her position. Trump could have offered no reason and been constitutionally entitled to act. However, in Comey's case, Trump had ample cause for firing him. Comey's direct supervisor was the deputy attorney general, Rod Rosenstein. As explained in the last chapter, Rosenstein spelled out several incidents of serious malfeasance by Comey in his mishandling of the Clinton email case that more than justified the decision, and the deputy AG recommended to the president that the director be dismissed.[33] So, too, did the attorney general, Jeff Sessions, who endorsed the decision.[34] Trump cited these reasons and recommendations in his letter to Comey, dated May 9, 2017.[35]

Rosenstein's endorsement that Comey be fired had been months in the making. In his testimony before both the House and Senate, he disclosed that he first broached the idea in a discussion with then-senator Jeff Sessions in the winter of 2016–17.[36] Hence, it was not a judgment motivated by some nefarious desire to interfere in the Flynn investigation or the Russia probe. The plan had been gestating for several months and predated Trump's conversation with Comey about Flynn. Instead, relieving the director of his duties was an effort "to restore the credibility of the FBI" because, as Rosenstein described it, Comey had "damaged public confidence in the bureau and department."[37]

The ink was barely dry on Trump's letter firing Comey when the president's critics began assailing the move as a clear case of obstruction of justice. Trump, they proclaimed, was interfering with the Trump-Russia investigation. This was untrue, and the facts do not support the argument. Not only was the president empowered to make such a decision and did so based on good cause, but the testimony of key people confirmed there was no such interference.

In a hearing before the Senate Intelligence Committee after Comey was fired, Acting FBI Director Andrew McCabe stated, "There has been no effort to impede our investigation to date."[38] Days later, Deputy Attorney General Rod Rosenstein told Congress, "There never has been . . . any political interference in any matter under my supervision in the United States Department of Justice."[39]

Comey himself disabused the notion that anyone had attempted to interfere in an FBI investigation when he responded this way to a general question about undue influence while testifying before Senate Judiciary Committee on May 3, 2017:

> I'm talking about a situation where we were told to stop something for a political reason, that would be a very big deal. It's not happened in my experience.[40]

That was Comey's story *before* he was fired. *After* he was dismissed, he changed his tune and testified on June 8, that he "understood the President to be requesting that we drop any investigation of Flynn" during the February 14 Oval Office meeting.[41] So which was the truth and which was a lie? It is likely that Comey's first account, when he was still the director, was the more accurate one. Had the president tried to obstruct the Flynn case months earlier, surely Comey would have said so when first questioned by Congress about any interference in FBI investigations. Only after he was fired, as an aggrieved or disgruntled ex-employee, did Comey have a motive to smear Trump by suggesting that the president had tried to impede the probe. But how would that even be possible? Trump fired Comey, not the entire FBI. Their investigation continued regardless of who was at the helm as director.

The subject of potential obstruction was also raised in a June 2017 Senate Intelligence hearing where other Trump administration officials testified. In the wake of a *Washington Post* report that claimed the president tried to intervene in the Russia investigation, the director of national intelligence, Dan Coats, said, "I have never felt pressure to intervene or interfere in any way with shaping intelligence in a political way or in relation to an ongoing investigation."[42] National Security Agency director Mike Rogers offered similar testimony. "In the three-plus years that I have been the director of the NSA, to the best of my recollection I have never been directed to do anything I believe to be illegal," Rogers said. "I do not recall ever feeling pressured to do so."[43]

Following Comey's firing, the president sat for an interview with NBC's Lester Holt. During the questioning, Trump referred to the Russia probe. Instantly, the president's opponents seized upon the televised interview as irrefutable proof of obstruction. It was not. A rigorous reading of what Trump said confirms that his intent was not to interfere with or end the Russia investigation, but to place someone who was neutral and competent in charge, given that Comey had committed multiple acts of wrongdoing in his handling of the Clinton case and had also refused to tell the public the truth that Trump was not personally under investigation.

Trump emphasized that he intended to fire Comey, "regardless of the recommendation" by Rosenstein and Sessions, although their guidance and endorsement was influential:[44]

> **TRUMP**: And in fact, when I decided to just do it, I said to myself—I said, you know, this Russia thing with Trump and Russia is a made-up story. It's an excuse by the Democrats for having lost an election that they should've won.
> **HOLT**: But were—are you angry with Mr. Comey because of his Russia investigation?
> **TRUMP**: I just want somebody that's competent. I am a big fan of the FBI. I love the FBI.[45]

Trump went on to explain that, "as far as I'm concerned, I want that thing to be absolutely done properly."[46] He then suggested that an even longer investigation might be appropriate to uncover all the facts:

> **TRUMP**: But I said to myself I might even lengthen out the investigation. But I have to do the right thing for the American people. He's (Comey) the wrong man for that position.[47]

As the interview continued, Trump expressed his frustration that Comey had repeatedly told him that he was not personally under investigation, yet refused to make that information public by telling the truth. According to the president, Comey "said it once at dinner and then said it twice during phone calls."[48] Despite Comey's reluctance to be honest with the American people, Trump made it clear that he never sought to halt the investigation:

> **HOLT**: Did you ask him to drop the investigation?
> **TRUMP**: No, never.
> **HOLT**: Did anyone from the White House ask him to end the investigation?
> **TRUMP**: No, no. Why would we do that?[49]

The president reiterated that Comey's earlier malfeasance called into question his ability to lead any significant probe, as well as the FBI itself. "I also want to have a really competent, capable director. He's not. He's a showboater," said Trump.[50]

The president insisted he did not ask that the Russia case be dropped and, to date, there is no evidence to the contrary.

All of this invites the question, was there anything damaging in Comey's memos? If these government documents, taken from FBI headquarters by Comey, were the foundation for any claim of obstruction, what exactly did they contain?

For the better part of a year, the FBI and Department of Justice guarded the memos with the utmost secrecy, refusing to release them to Congress. Finally, under threat of contempt, the documents were handed over on April 19, 2018, and immediately made available to the public. The memos were notable for what they did *not* contain. Although Comey was quite detailed about what occurred during his meetings with the president, not once did he claim that Trump had tried to obstruct the FBI's investigation. To the contrary, in one memo dated March 30, 2017, the president encouraged Comey to *continue* his investigation to determine whether anyone in his campaign collaborated with the Russians because "it would be good to find that out."[51]

So where was the evidence of obstruction? If it was not found in Comey's stolen memos and was not discerned from Trump's interview with NBC, how could it ever be shown, much less proven in a court, that he harbored a "corrupt" or "improper purpose," as the law of obstruction demands? The answer is that no such evidence existed. None. Like the fabricated argument of Trump-Russia "collusion," there was no case for obstruction of justice.

If Comey is the key witness in any obstruction case, then the following exchange during his Senate Intelligence Committee hearing is likely the best evidence that neither Trump nor anyone else attempted to impede the Russian investigation:

> **SEN. BURR**: Did the president at any time ask you to stop the FBI investigation into Russian involvement in the 2016 U.S. elections?
>
> **COMEY**: Not to my understanding, no.

SEN. BURR: Did any individual working for this administration, including the justice department, ask you to stop the Russian investigation?
COMEY: No.[52]

Some have suggested that comments made by the president to visiting Russian officials in the Oval Office the day after Comey was fired might somehow serve as evidence of obstruction. They do not. The *New York Times* reported that when Foreign Minister Sergey Lavrov and Ambassador Sergey Kislyak came calling, Trump allegedly quipped, "I just fired the head of the FBI. He was a crazy, a real nut job. I faced great pressure because of Russia. That's taken off."[53]

The statement, assuming it was accurately reported, does not prove a "corrupt" intent to dismiss Comey for the hidden purpose of halting his investigation. More plausibly, it is evidence that Trump felt pleased and optimistic that someone else would be in charge of the case who could be objective, unbiased, and "competent."

In a desperate search for some other act of obstruction, the *Los Angeles Times* reported that a statement issued to the media by Donald Trump Jr. describing his Trump Tower meeting with a Russian lawyer during the campaign was untruthful and, therefore, it was being examined by the special counsel as evidence of obstruction.[54] This is patently ridiculous. Let's assume the media statement did not accurately describe the full extent of the meeting. Let's assume it was a bald-faced lie. Lying to journalists is not, and never has been, a crime. If it were otherwise, just about every politician in Washington, D.C., would be behind bars.

However, the statement describing the meeting at Trump Tower was, at its core, accurate. Here is the relevant part:

> We primarily discussed a program about the adoption of Russian children that was active and popular with American families years ago and was since ended by the Russian government, but it was not a campaign issue at the time and there was no follow up. I was asked to attend the meeting by an acquaintance, but was not told the name of the person I would be meeting with beforehand.[55]

Both the Russian lawyer and Trump Jr. have stated that the meeting, scheduled under false pretenses, was a conversation about Russian adoptions and little more. A subsequent statement issued by the president's son explained that the lawyer claimed to have information "that individuals connected to Russia were funding the Democratic National Committee and supporting Ms. Clinton." But, Trump Jr. added, "her statements were vague, ambiguous and made no sense. No details or supporting information was provided or even offered."[56] Moreover, in an interview with Fox News, the Russian lawyer insisted there was no discussion whatsoever about Clinton, and she had no such damaging information about the candidate.[57]

Whether President Trump dictated the press statement or participated in its composition is of no legal consequence. Even if it was an attempt to cover up an effort to deceive the media, that is not a crime. The fact that a revised statement was later issued providing a more extensive and complete account of the Trump Tower meeting proves nothing other than a desire to be more forthcoming and transparent. It is not an admission that the original press release was evasive or deceitful.

There is one more argument that militates against any claim of obstruction in the firing of Comey or any other of the actions mentioned herein. It is hard to make a case for obstructing a noncrime.

As a legal matter, it can be done with some jurisprudential gyrations. But from a practical and reasoned standpoint, it is difficult for a prosecutor to argue that the president intended to obstruct an inquiry that he felt confident would prove his innocence. With no evidence that he ever colluded with the Russians during the campaign, the investigation of Trump was entirely without merit. If the president knew he did nothing wrong, it makes little sense that he would try to impede a probe into no wrongdoing.

It is clear that Trump felt falsely accused and wrongfully convicted in the court of public opinion. Democrats and the liberal media were relentless in banging the daily drum that Trump had engaged in all manner of illegality, despite a paucity of evidence. Beleaguered and under siege, the president implored Comey to tell the public the same truth he had been told in private—that he was not under criminal investigation. Comey, who probably thought he had the upper hand, refused to lift the black veil of suspicion. He failed to do the right thing. It is no wonder the president became exasperated with the FBI director and chose to find a replacement who could be honest, competent, and fair.

Former federal prosecutor Doug Burns is convinced there were ample grounds to fire Comey, none of which amounted to obstruction of justice:

> Comey's announcement of a no prosecution of Hillary Clinton was totally improper, as FBI officials or agents never make such prosecutorial decisions. Similarly, Comey's factual condemnation of her was totally inappropriate. And last, his explanations of the law were both bizarre and incorrect. While the president did himself no favors in his interview with NBC's Lest Holt, legally the Comey termination was not obstruction of justice.[58]

While no legitimate case can be made against Trump for obstruction, a case *can* be made that Comey committed crimes.

COMEY'S THEFT OF GOVERNMENT DOCUMENTS

For years, James Comey carefully cultivated a public portrait of himself as a grown-up Boy Scout—honest and morally straight. The truth is quite different. His actions belie the virtuous image he sought. It was all an illusion designed to mask the kind of conduct that most people find reprehensible. The record shows that he was less than honest, engaged in acts of questionable legality, and abused his power to further his ambitions.

One of the more stunning moments during Comey's testimony before the Senate Intelligence Committee in June 2017 occurred when he confessed that he deliberately leaked to "a friend" the contents of the presidential memorandums memorializing his conversations with Trump.[59] He directed that friend, Daniel Richman of Columbia Law School, to leak the information to the *New York Times* with the objective that it would trigger the appointment of a special counsel to investigate the man who had just fired him. It was a devious scheme, to be sure. Comey *knew* the media would be more than willing to trash Trump by contorting the memos' contents and misconstruing the law to accuse the president of obstruction of justice. Journalists and pundits did not disappoint.

The opening sentence in the *Times* story on May 16, 2017, did not recite facts derived from the memos, but drew an unsupported conclusion that "President Trump asked the FBI director, James Comey, to shut down the federal investigation into Mr. Trump's former national security adviser, Michael T. Flynn, in an Oval Office meeting in February."[60] Ipso facto, obstruction. The headline was nearly identical to the first line. Thus, anyone who did not read past the title of the story or the opening sentence was led to believe that Trump had probably committed a crime.

Of course, this is not what happened in the February meeting, according to Comey, who testified on June 8 about his conversation with Trump, narrating the encounter from his memos. Indeed, at the congressional hearing, Comey specifically quoted Trump's vague comments about Flynn as "hoping" he would be cleared.[61] That is not the same thing as "asking to shut down" an investigation, as the *Times* would have its readers believe. The *Times* story went on to raise the specter of obstruction and, sure enough, the next day Comey's longtime friend and ally Robert Mueller was appointed special counsel. For the fired FBI director, it was mission accomplished. His media leak achieved his desired purpose.

In defense of his actions, Comey offered an explanation that was, in equal parts, erroneous and obtuse. He claimed that the seven presidential memos he took with him when he was fired were his personal property. If he believed that, he is not much of a lawyer. The FBI's policy manual states quite clearly that documents or records generated during official duties are government property.[62] The FBI Employment Agreement, to which Comey was bound, mandates that "all information acquired by me in connection with my official duties . . . remain the property of the United States of America."[63]

Under both the Federal Records Act and the Privacy Act, any document or record composed by government employees during the course and scope of their employment is not the property of the person who authored the document, but the property of the government.[64] This is especially true if the material was prepared on a government-owned computer and written during the normal work hours while the employee is on the job performing the duties of his job, as was the case with Comey's presidential memos. His discussions with the president arose directly from his position as head of the FBI. These records laws apply to classified and unclassified documents alike. Furthermore, in his testimony before the Senate Intelligence Committee on June 8, 2017, Comey admitted that he wrote the memos so that they could be "discuss(ed) within the FBI and the government."[65] This is an admission that these documents were not his personal property. Records that are composed for government use are automatically government property.

The fact that Comey did not want to leak the memos himself, but chose a conduit or middleman to do so covertly at his behest, is substantial proof that he knew what he was doing was wrong and illegal. By using a third-party to do the dirty work, Comey was trying to circumvent the law to insulate himself from criminality. He failed.

18 U.S.C. 641 makes it a felony punishable by up to ten years in prison to give someone outside of government an unclassified, but protected, record without authorization:

> Whoever embezzles, steals, purloins, or knowingly converts to his use or the use of another, or without authority, sells, conveys or disposes of any record . . . of the United States or of any department or agency thereof . . . shall be fined under this title or imprisoned not more than ten years.[66]

This is precisely what Comey appeared to have done "converting" to his own use and in conveying to his friend, without authorization, his presidential memos which were government records.

Having been fired, Comey stole government records with the intent to leak them for his benefit. In an obvious act of retribution, he wanted the documents to inculpate Trump in a special counsel investigation and, he hoped, generate a criminal charge of obstruction of justice. This scheme to benefit himself and harm the president also may have violated at least two federal regulations, including this one identified in the Code of Federal Regulations:

> An employee shall not engage in a financial transaction using nonpublic information, nor allow the improper use of nonpublic information to further his own private interest or that of another, whether through advice or recommendation, or by knowing unauthorized disclosure.[67]

Under the law, it does not matter that Comey was an ex-employee when he leaked the documents because he maintained custody of them when he was still employed, then took them out of the FBI building to use for his own devices. This was a direct violation of FBI regulations which state, "FBI personnel must surrender all materials in their possession that contain FBI information upon FBI demand or upon separation from the FBI."[68] Comey did not do this. He converted government property to his own use, then disseminated it to the public.

Comey must have known that he was likely breaking several laws and committing felonies. As FBI director, he was legally obligated to adhere to the bureau's standard nondisclosure contract in which all personnel promise not to disclose the very type of records and information Comey leaked. The agreement specifically warned that employees are subject to "criminal sanctions . . . and personal liability in a civil action at law . . . and the disgorging of any profits arising from any unauthorized publication or disclosure."[69] Separation upon termination did not render the contract null and void. It was a binding, enforceable, and actionable contract regardless of job status. Under the terms, Comey agreed he could be sued and face criminal prosecution. Since his firing, Comey published a book quoting from the memos he filched. This enabled him to profit handsomely from his wrongful actions, pocketing millions of dollars. If the FBI contract were to be enforced, Comey could—and should—lose earnings derived therefrom.

The Comey-composed memos themselves recited discussions with the president that were both privileged and contained information involving an ongoing FBI investigation into Flynn's contacts with Russia. This means Comey appeared to have broken yet another law, punishable by up to ten years in prison. 18 U.S.C. 793 makes it a crime to "willfully communicate or transmit national defense information," even though it is not necessarily classified when written.[70] While the full contents of the partially redacted memos made public so far do not deal directly with national defense matters, the overall Flynn investigation did.

Comey's chicanery was laid bare in his congressional testimony when he told the Senate Intelligence Committee that he deliberately wrote some of his memos as "unclassified" documents. Making them classified, he told the committee, "would tangle them up."[71] In other words, he manipulated the classification system to exploit the political damage his documents might cause, while concomitantly attempting to shield himself from criminal charges. But this may be a moot point if any of the seven memos Comey took with him *contained* classified information, regardless of how he might have labeled them or, more aptly, mislabeled them. Under law, the *content* dictates classification, not the markings.

Sometime in late 2017 or early 2018, the FBI advised the Senate Judiciary Committee that the majority of the memos were, in fact, "classified."[72] Chairman Charles Grassley, one of the few people who gained access to the memos, revealed that four of them were "marked classified at the 'Secret' or 'Confidential' levels," a fact that was confirmed when the memos were released.[73] Richman told Fox News that he received four of the seven memos.[74] This means that Comey appears to have given his "friend" at least one "classified" document.

Giving "classified" records to an unauthorized person and/or storing them in an unsecured venue constitutes several felonies—the same crimes Hillary Clinton surely committed. For example, 18 U.S.C. 1924 states as follows:

> Whoever, being an officer, employee of the United States . . . becomes possessed of documents or materials containing classified information of the United States, knowingly removes such documents or materials without authority and with the intent to retain such documents or materials at an unauthorized location shall be fined under this title or imprisoned for not more than five years, or both.[75]

Comey appears to have done this. He admitted he knowingly removed presidential memos without authority from FBI headquarters, kept them in what must have been an unauthorized location, then conveyed at least four of them to his "friend," Richman. As director of the FBI, he knew that at least some of their contents were both privileged and might well be classified. It would be folly for Comey to argue they were not classified since the FBI insists they are. If Comey deliberately mismarked them, he cannot use his own wrongful act to insulate himself from criminal prosecution.

In the alternative, let's assume for the sake of argument that Comey's handling of the documents was "grossly negligent," instead of "knowing" or "intentional." That would be the same crime for which Clinton should have been charged, 18 U.S.C. 793(f).[76] The irony is lost on no one. Comey appears to have committed the identical felony as Clinton, and it was Comey who contorted the law to absolve her of this crime, as explained in chapter 2.

But the story of Comey's machinations does not end there. Days after the presidential memos were released to the public, it was learned that Richman had worked for Comey at the FBI as an unpaid "special government employee."[77] Comey concealed this important information from Congress during his June 2017 testimony, later dismissing this fact as "irrelevant."[78] Moreover, Comey failed to disclose that another person, Patrick Fitzgerald, also reportedly received memos.[79] Fitzgerald is a former U.S. Attorney and special counsel who, like Richman, is a friend of Comey. Both Richman and Fitzgerald have since been hired by Comey as his lawyers.[80] And so, too, has another lawyer, David Kelly, to whom Comey gave one or more memos.[81] This means that the fired director can invoke the attorney-client privilege to try to protect some or all of their communications about the memos.

The FBI was sufficiently concerned about Comey's dissemination of classified information that agents conducted a search of Richman's office to retrieve documents and contain the leak.[82] It is unknown whether the same "spillage clean-up" occurred at Fitzgerald's office and, perhaps, Kelly's, as well. These corrective actions by the FBI suggest that classified information may well have been shared by Comey in violation of federal law.

When Comey was questioned by senators in a June 2017 hearing before the Intelligence Committee, he omitted these relevant and important details in his answers about his leak of the memos. Under 18 U.S.C. 1001, it is a crime to give false or misleading statements in a legislative proceeding.[83] "Concealing material facts" in response to questions under oath before Congress would constitute misleading statements in violation of that statute.

Congress has been investigating Comey for a series of other

suspected deceptions made during testimony before various congressional committees. In one instance, he told the House Judiciary Committee, under oath, that he decided not to refer criminal prosecution of Clinton only *after* she was interviewed.[84] Yet, documents uncovered later indicated he made the decision well *before* the interview.[85]

Comey insisted that Loretta Lynch, the attorney general, never knew of his decision to clear Clinton in advance of his public announcement.[86] Yet, text messages exchanged between Peter Strzok and Lisa Page suggested that Lynch had been apprised in advance.[87] Comey also testified that, while FBI director, he never authorized leaks to the media about the two presidential candidates.[88] Yet, a subsequent statement by his deputy director, Andrew McCabe, appeared to contradict Comey.[89]

Finally, the Senate Judiciary Committee sent a letter to the Justice Department's inspector general accusing Comey of "apparent material discrepancies" in his testimony about the FISA warrant applications, asking whether this was "a deliberate attempt to mislead."[90]

There is substantial evidence that Comey did not tell the truth on several occasions and may have violated numerous federal statutes governing the theft of government documents, including classified material. He may also have obstructed justice in the Hillary Clinton email case and violated the law by deceiving the FISA court in a warrant to spy on an American citizen.

Days after Comey published his book and commenced his publicity tour, it was learned that the inspector general at the Department of Justice was investigating whether Comey mishandled classified information contained in the presidential memos he gave to his "friend" that was then leaked to the media.[91] If he broke the law, he should be held accountable.

Former independent counsel and U.S. Attorney Joe diGenova was blunt in his assessment of Comey:

> I don't think there's any doubt that Comey committed multiple crimes. If the Justice Department wants to pursue them vigorously and fairly like they would with any other citizen, he should be indicted for his false testimony on Capitol Hill and for his obstruction of an investigation.[92]

Far from the image of an honest and honorable Boy Scout, the evidence is compelling that James Comey sought to mislead, deflect, and deceive. He also appears to have abused the powers of his office to exact punishment on the president who fired him. His plan to convert presidential memos for his own use, then leak them to the media to damage Trump suggests a willingness to defy rules, regulations, and federal laws with impunity.

Perhaps Comey felt he could get away with it because he successfully engineered the appointment of his close friend Robert Mueller as special counsel to pursue potential charges against the president.

CHAPTER 11

THE ILLEGITIMATE APPOINTMENT OF ROBERT MUELLER

The primary duty of a lawyer engaged in public prosecution is not to convict, but to see that justice is done.

—*Canons of Professional Ethics*, Canon 5 (1908)

Ages ago, the legendary criminal defense attorney Clarence Darrow gazed back on his long career as a central figure inside America's courtrooms and sadly observed to a reporter, "There is no such thing as justice—in or out of court."[1]

As a young lawyer filled with idealism, I thought he was wrong. Nearly four decades later, I fear he was right. I was naïve. Darrow knew through experience that, all too often, prosecutors are willing to abandon their ethical duty to see that justice is done. Instead, emboldened by their powers, they become consumed by a zeal to win, sometimes at any cost. Fairness and justice are secondary to the goal of gaining a conviction. The law, Darrow concluded, had evolved into a "horrible business."[2]

Exhibit A in support of Darrow's thesis is the appointment of Robert Mueller to serve as special counsel. The facts and circumstances did not warrant a special counsel. The regulations governing his selection were misused. His authority exceeded the limits intended. And Mueller, himself, was so stricken with conflicts of interests that he should never have accepted the position. His determination to forge ahead, notwithstanding these legal obstacles, demonstrates that considerations of justice were a mere nuisance, if not irrelevant.

To put it as succinctly as possible, Mueller launched an investigation in search of a crime—any crime. This is backwards. Under regulations authorized by law, there must first be an identifiable crime specifically stated and accompanied by some evidence that such a crime may have been committed by someone. Mueller would then be empowered to question witnesses, examine records, and gather documents to resolve whether any charges should be brought. However, this legal requirement was not met when the special counsel was appointed. This makes his endeavor a lawless or illegitimate one.

Mueller has a long track record of loyal service to our nation, and many trust him to be fair and impartial, as if that excuses an improper appointment. The team of lawyers he hand-picked to assist him is as riddled with conflicts of interest as he is.

MUELLER'S APPOINTMENT WAS INVALID

After taking office as attorney general, Jeff Sessions incorrectly disqualified himself from any investigations involving Russia and the 2016 campaign. He was not required to do so, as detailed in chapter 9. However, his mistaken recusal meant that the deputy attorney general, Rod Rosenstein, became "acting attorney general" on all matters involving Sessions's disqualification.

On May 17, 2017, Rosenstein decided to appoint Robert Mueller as special counsel. The authority to make an appointment is derived from federal statutory law (28 U.S.C. 515), and defined by the Code of Federal Regulations, 28 C.F.R. 600.1 which states:

> The Attorney General, or in cases in which the Attorney General is recused, the Acting Attorney General, will appoint a Special Counsel when he or she determines that criminal investigation of a person or matter is warranted.[3]

These regulations are codified into law known as administrative laws. All government employees, regardless of rank or position, are required to follow and uphold them. This particular regulation makes specific reference to a "criminal investigation." To conduct one, there must first be some evidence that a *crime* might have been committed to warrant the investigation. When Mueller was appointed, there was no suspected crime. Moreover, no other campaign-related statutes had any relevance to the known facts involving activities in the Trump campaign.

The importance of explicitly stating a suspected crime as a necessary predicate for appointing a special counsel is reinforced by the requirement set forth in 28 C.F.R. 600.4:

> The Special Counsel will be provided with a specific factual statement of the matter to be investigated.[4]

Rosenstein's appointment order provided no such "specific factual statement" of a matter to be criminally investigated. Instead, the order granted Mueller broad authority to investigate matters that are not criminal at all. The FBI's probe up to that point was a "counterintelligence investigation," according to Director James Comey's testimony before the House Intelligence Committee.[5] The bureau was looking into Russian interference in the 2016 election, a national security concern. Information was being collected. That is not the same as a criminal investigation. A strong case can be made that Mueller's appointment was defective and invalid because he was empowered to investigate something that was not authorized by the special counsel regulation.

Former federal prosecutor Andrew McCarthy argued that Mueller's appointment violated the special counsel regulation because "the regulation does not permit the Justice Department to appoint a special counsel in order to determine whether there is a basis for a criminal investigation. To the contrary, the basis for a criminal investigation must pre-exist the appointment." He added, "It is the criminal investigation that triggers the special counsel, not the other way around."[6] He is correct. A close examination of the order appointing Mueller shows that the Justice Department reversed the process and did it backward.

There are three parts to Rosenstein's appointment order. They are worth examining separately. The *first* part reads as follows:

> The Special Counsel is authorized to conduct the investigation confirmed by then-FBI director James B. Comey in testimony before the House Permanent Select Committee on Intelligence on March 20, 2017, including . . . any links and/or coordination between the Russian government and individuals associated with the campaign of President Donald Trump.[7]

Where was the stated crime? The "specific factual statement" was not specific in any sense of the word. It was quite broad and open-ended. It referenced Comey's testimony in which he had revealed an outstanding counterintelligence investigation, but not a criminal case. Contrary to the requirements of the special counsel regulations, Rosenstein's order called for an investigation of something that was not illegal. Assuming there was ever any evidence of "links" or "coordination" between the Russians and the Trump campaign, where in the criminal codes does that rise to the level of criminal activity? It does not. Perhaps someday Congress will decide to make it a crime. Until then, it is perfectly legal.

Michael B. Mukasey, former U.S. attorney general and chief judge of the United States District Court for New York, pointed out this legal defect in Mueller's appointment order when he wrote a column for the *Wall Street Journal:*

> Possible Russian meddling in the 2016 election is certainly a worthy subject for a national-security investigation, but "links" or "coordination"—or "collusion," a word that does not appear in the letter of appointment but has been used as a synonym for coordination—does not define or constitute a crime.[8]

Mukasey noted that prior investigations utilizing independent counsels such as Watergate, Iran-Contra, and Whitewater "identified specific crimes" and "the public knew what was being investigated."[9] The whole point of the special counsel regulation was to avoid limitless investigations by unaccountable prosecutors and to apprise the public of what potential crimes were being pursued. The public, after all, has a right to know if their tax dollars are being used or misused for an unbounded search for unidentified crimes.

But with Mueller's appointment, Americans were given no idea what the special counsel was up to because Rosenstein's assignment order violated the regulation that demanded the identity of a specific crime to be investigated. Not only was Mueller's appointment invalid for this reason, but there was no need for a special counsel at all, in the absence of some evidence of a crime. The FBI could have continued its counterintelligence probe.

In an interview with the Public Broadcast Network, Mukasey elaborated:

> Any links or coordination may sound sinister, but it doesn't define or suggest the existence of a crime. So, what's been defined here, or not defined here, is something that looks very much like a fishing expedition or a safari.[10]

This brings us to the *second* part of Rosenstein's order that directed Mueller to investigate "any matters that arose or may arise directly from the investigation."[11] This was an even bigger blank check and completely in contravention of the special counsel regulation. Rosenstein had no authority to confer such an expansive and undefined mandate for Mueller because, again, the law demands a "specific factual statement of the matter to be investigated." He is not permitted to say simply, "and look for anything else that might come up." This flaw or defect in the order also rendered Mueller's appointment invalid. He cannot exercise power that Rosenstein had no authority to grant.

The *third* part of Rosenstein's order instructed Mueller to investigate "any other matters within the scope of 28 C.F.R. 600.4(a)."[12] What does that regulation state? It reads:

> The jurisdiction of a Special Counsel shall also include the authority to investigate and prosecute federal crimes committed in the course of, and with intent to interfere with, the Special Counsel's investigation, such as perjury, obstruction of justice, destruction of evidence, and intimidation of witnesses.[13]

In simple language, this means that if someone tried to interfere with Mueller's investigation or lied during questioning, he or she could be prosecuted for obstruction of justice, perjury, or making a false statement. This did not grant the special counsel authority to investigate Trump for any obstruction of justice that might have happened *before* the special counsel was appointed. Only impeding Mueller's investigation, itself, is recognized by the regulation.

So, what was Mueller left with? Investigating any "links or coordination" with the Russians. But since that is not a crime in a political campaign, then Mueller was tasked with investigating a noncrime which he is not authorized to do under the special counsel regulation. He had no legitimate assignment and no regulatory authority to do anything.

Rosenstein and Mueller likely *knew* they were bending the law into a pretzel. When the special counsel indicted Paul Manafort on charges of tax, financial, and bank fraud crimes that had nothing to do with the Trump campaign and significantly pre-dated his brief role as campaign manager, his lawyers filed a motion to dismiss the case by challenging Mueller's authority to prosecute based on some of the arguments stated herein.

In response, the special counsel filed documents with the federal court *admitting* that Rosenstein's order appointing Mueller was *deliberately vague*. This was an astonishing confession that the acting attorney general had carefully crafted his appointment order in a way that willfully violated the special counsel regulation that required a specific statement of facts. Rosenstein had no intention of adhering to the law when he named Mueller to the post on May 17, 2017.

The court documents revealed that Rosenstein issued a new, *secret* memorandum to Mueller, dated August 2, 2017. In it, Rosenstein seemed to concede that his original order appointing Mueller did not comply with the special counsel regulations:

> The May 17, 2017, order was worded categorically in order to permit its public release without confirming specific investigations involving specific individuals.[14]

Rosenstein's memo then proceeded to order Mueller to investigate "allegations that Paul Manafort committed a crime or crimes by colluding with Russian government officials with respect to the Russian government's efforts to interfere with the 2016 election for President of the United States, in violation of United States law."[15] What law? Conspicuously, Rosenstein failed to cite a law or potential crime, perhaps having difficulty in finding one that had any application to "collusion." Therefore, even the second secret order failed to meet the requirement of the regulation that a crime be stated.

Importantly, this memo was hidden from the public and only disclosed because Mueller was forced to do so when faced with the prospect that a federal judge might dismiss the case on the grounds that Mueller's appointment was invalid and he, therefore, lacked any authority. Had it not been for Manafort's lawyers who contested the basis for the special counsel's appointment, Rosenstein's memo would still be a secret. They were misleading Americans. They made available an opaque order for public consumption, but held back a different order containing the truth.

It seems apparent that Rosenstein and Mueller conspired to circumvent the special counsel regulation. In May, Rosenstein knowingly wrote a vaguely worded order when the law demanded a specific one. The two men then decided between themselves what and whom to investigate and to keep it concealed from the American public. This violated both the letter and spirit of the special counsel regulations. As written, the rules promised clarity and transparency. Yet, Rosenstein and Mueller worked in concert to undermine them. In so doing, they have further eroded the public's trust in the fairness and integrity of the legal process.

Their questionable conduct cannot be cured by their belated effort to comply with the strict requirements of the regulation. Why? Because they engaged in still another deception. Note the date of the subsequent memo authorizing Mueller to investigate Manafort—August 2. However, Mueller ordered the FBI to conduct a predawn raid of Manafort's home on July 26, a full week before the issuance of Rosenstein's order. Since Mueller claimed that the August 2 memo authorized him to pursue Manafort, this was an admission that he was *not* authorized at the time of the raid. Arguably, any material illegally seized would be inadmissible evidence under the well-established "exclusionary rule."[16]

Another curious anomaly jumps out from Rosenstein's August 2 memo where he wrote that Mueller's authority to investigate Manafort was "within the scope of the investigation at the time of your appointment and are within the scope of the Order."[17] Why, then, was there a need to write the second memo at all? And why did it take nearly three months to do it? The obvious answer is that Rosenstein and Mueller must have recognized that they had not complied with the regulations and were endeavoring to correct their mistake retroactively.

Even the second memo failed to remedy the problem, since it

also neglected to specify what federal laws might have been violated as the basis for the criminal investigation. There must first be a criminal investigation before a special counsel can be named. This did not exist. There was only a counterintelligence probe at the time of Mueller's appointment. That probe involved the gathering of information. The special counsel regulations do not allow an appointment to be used for this purpose. If it were otherwise, there would be an endless number of special counsels bootstrapped to all the various FBI counterintelligence cases that involve Russia and from which Sessions might be recused. Harvard law professor Alan Dershowitz concluded that both of Rosenstein's orders failed to comply with the special counsel regulations: "You can't investigate sins, you can only investigate federal crimes and there is no such federal crime as 'collusion.'"[18]

In court documents, Mueller argued that his original authorization order was deliberately vague because anything specific would be confidential and might jeopardize his investigation.[19] This is a standard excuse by prosecutors. The more plausible explanation is that the August 2 memo was a delinquent effort by Rosenstein and Mueller to remedy their failure to comport with the regulations. They surely knew they had violated these special counsel rules, made worse by their raid of Manafort's home. So, they attempted to rectify their mistake after the fact by devising a new, albeit equally defective, order. Then, under the guise of "confidential," they appeared to have covered it up to conceal from the public what they had done.

None of the charges against Manafort are even remotely related to any "coordination" between the Trump campaign and Russia during the election nor "any matters that arose or may arise directly from the investigation," as the Mueller's Order of Appointment specifically authorizes.[20] Nowhere is Manafort accused of crimes involving Russian officials or their government, and the special counsel has provided no evidence of it.[21] Instead, the criminal complaint alleges wrongful conduct between 2006 and 2014. This significantly predates the presidential election. Moreover, it appears that Mueller did not uncover new evidence during the course of his investigation, but resurrected an old Justice Department investigation of Manafort in which no charges were ever brought. In federal court, lawyers for the special counsel admitted it. Judge T.S. Ellis III then accused Mueller's team of exerting "unfettered power" to bring down the president:[22]

> I don't see what relation this indictment has with anything the special counsel is authorized to investigate.
>
> Now, I think you've already conceded appropriately that this investigation that has led to this indictment long antedated the appointment of a special prosecutor; that it doesn't have anything to do with Russia or the campaign; and he's [Manafort] indicted; and it's useful, as in many cases by prosecutors, to exert leverage on a defendant so that the defendant will turn and provide information on what is really the focus of the special prosecutor.

As the judge correctly reasoned, Mueller is acting in violation of the special counsel regulations and abusing his authority.[23]

MUELLER SHOULD HAVE RECUSED HIMSELF

The appointment of Mueller originated when fired FBI director James Comey took presidential memos out of the FBI building, converted them to his own use, and then purposefully delivered them to a friend to leak to the media. Appearing before the Senate Intelligence Committee in June 2017, Comey admitted he engineered the leak because "I thought that might prompt the appointment of a special counsel."[24] It was a typical Comey maneuver—deviously effective.

What has never been explained is how one of Comey's closest friends in Washington, Mueller, ended up being selected as the special counsel. The decision was supposedly made by Rosenstein who has never fully explained what went on behind the scenes. But it is well known that Comey, Mueller, and Rosenstein were friends and colleagues in their various roles throughout the years with the Department of Justice. The fact that all three became involved in the investigation of Trump after the president fired one of them calls into serious question their objectivity. It is quite the cozy arrangement. As such, it is saturated with conflicts of interests.

The law governing the special counsel, 28 C.F.R. 600.7, prohibits someone from serving if he has a "conflict of interest."[25] The same Code of Federal Regulations defines what constitutes a conflict:

> No employee shall participate in a criminal investigation or prosecution if he has a personal or political relationship with:
>
> 1. Any person substantially involved in the conduct that is the subject of the investigation or prosecution; or
> 2. Any person which he knows has a specific and substantial interest that would be directly affected by the outcome of the investigation or prosecution.[26]

The regulation then explains that a personal relationship includes "friendships" and can be a cause for disqualification if they pose "a close and substantial connection of the type normally viewed as likely to induce partiality."[27] Several other regulations calling for recusal also govern the special counsel.

In applying these rules to the Mueller probe, it is clear that Comey is "substantially involved" in any obstruction investigation of Trump, as reported by the *Washington Post*.[28] Indeed, he appears to be the only witness to the alleged Oval Office discussion about Flynn and was involved in conversations with the president that led to his firing. Since he admits he sought a special counsel to investigate Trump, he obviously has a "substantial interest" in the "outcome" of Mueller's probe and potential prosecution. He is hardly an innocent or disinterested bystander.

A conflict of interest is a situation in which an individual has competing interests or loyalties. Here, it sets up a clash between the special counsel's self-interest or bias and his professional or public interest in discharging his responsibilities in a fair, objective, and impartial manner. Mueller's close friendship with the key witness raises the likelihood of prejudice or favoritism which is anathema to the fair administration of justice. How can Americans have confidence in the results if they suspect that the special counsel may harbor bias? They cannot. The conflict inevitably discredits whatever conclusion is reached. It renders the entire investigatory exercise suspect, and it only elevates the controversy surrounding it.

Mueller had no choice but to disqualify himself. The law afforded him no discretion because the recusal is mandatory in its language. It does not say "may" or "can" or "might." It says the special counsel "shall" recuse himself in such instances.

The Mueller-Comey friendship is well documented and indisputable. They have long been friends, allies, and partners. Their bond is driven by a mentor-protégé relationship which makes the likelihood of favoritism and partiality self-evident. This represents an acute conflict of interest. Even the *appearance* of a conflict merits recusal under the law (28 U.S.C. 528).[29] Rules of professional responsibility that govern lawyers also prohibit the appearance of conflicts.[30] Mueller might argue he believes he is capable of being neutral, but that is not the point of the law that requires disqualification based on appearances.

Mueller and Comey worked together in their respective positions at the FBI and Justice Department. In tandem, they handled several important cases. In one memorable case, they stood in solidarity, both threatening to resign over the warrantless wiretapping fiasco involving then-attorney general John Ashcroft in 2001.

But the Mueller-Comey duo is best known for "badly bungling the biggest case they ever handled," in a story recounted by journalist and author Carl M. Cannon, Washington bureau chief for RealClearPolitics.[31] Letters laced with anthrax killed five people and infected seventeen others in Washington in 2001. Mueller, Comey, and others misinterpreted the evidence and botched the case by blaming an innocent man, Steven Hatfill, principally because drug-sniffing dogs seemed to take a liking to the suspect who had petted them. Cannon described it this way:

> Mueller, who micromanaged the anthrax case and fell in love with the dubious dog evidence, personally assured Ashcroft and presumably George W. Bush that in Steven Hatfill the bureau had its man. Comey, in turn, was asked by a skeptical Deputy Secretary of Defense Paul Wolfowitz if Hatfill was another Richard Jewell—the security guard wrongly accused of the Atlanta Olympics bombing. Comey replied that he was "absolutely certain" they weren't making a mistake.

Such certitude seems to be Comey's default position in

his professional life. Mueller didn't exactly distinguish himself with contrition, either.[32]

Their mistake ended up costing taxpayers $5.82 million in a legal settlement. But here is the interesting part that few people recall: Hatfill's successful lawsuit accused the FBI and DOJ of leaking information about him to the media in violation of the Federal Privacy Act. Sound familiar? It was very much like Comey's premeditated leak to the media of his now infamous memo reciting his alleged conversation with Trump. Perhaps, old habits are hard to break.

Other news organizations have chronicled Mueller's close friendship, including the *Washington Post* which entitled its story, "Brothers in Arms: The Long Friendship Between Mueller and Comey."[33] Perhaps most revealing was a lengthy *Washingtonian* story in 2013 describing in detail a deep friendship that stretches back over a decade. Mueller and Comey became "close partners and close allies." So close, that at one critical juncture, "there was only one person in government whom Comey could confide in and trust: Bob Mueller."[34]

Against this backdrop, the inevitable conflict of interest comes into full view. If the special counsel is investigating whether the president tried to obstruct justice, the case becomes a test of "he said . . . he said." Which man will Mueller believe? His good friend? Or the man who fired his good friend? How can Mueller fairly and impartially assess Comey's credibility versus Trump's? He cannot.

Equally important, how can the public be assured that Mueller's decision is free of prejudice, if not animus, driven by his affection for Comey? It is reasonable to assume that Mueller was not pleased to see his good friend fired by Trump. Might the special counsel be tempted to exact retribution by conjuring criminality or other wrongdoing where none may exist?

There have been innumerable calls for the special counsel to recuse himself in favor of someone who could be more objective. Prominent law professors penned columns in *USA Today* requesting Mueller's resignation.[35] The editorial board of the *Wall Street Journal* emphasized how Mueller spent months hiding from Congress the truth about why Peter Strzok and Lisa Page had been removed from the special counsel probe. The board urged Mueller to step aside, noting that "the investigation would surely continue, though perhaps with someone who doesn't think his job includes protecting the FBI and Mr. Comey from answering questions about their role in the 2016 election."

Mueller's conflict of interest is twofold. It extended beyond his fast friendship with Comey, a pivotal witness in any obstruction case. Noted George Washington University legal scholar Jonathan Turley argued that Mueller's meeting with the president created an additional disqualifying conflict:

> The White House has said that Mueller interviewed with Trump for the FBI position to replace Comey the day before he was made special counsel. If this conversation occurred, Trump may have explained why he fired Comey and what he was looking for in his replacement. With obstruction a focus of the investigation, that makes Mueller a possible witness in his own investigation.[36]

In other words, on a Tuesday Mueller was interviewed by the president to become the FBI director, but on Wednesday Mueller accepted the job to investigate the president. Surely, Mueller knew he was under consideration to be named special counsel. Such important appointments don't unfold spontaneously in a day. Still, as he was sitting there talking with Trump, he never advised the president that he might investigate him. Was Mueller secretly attempting to gather evidence from Trump to incriminate him? Did they discuss the president's reasons for firing Comey which Mueller could then use in an obstruction case against Trump? Even if Mueller is not a witness in his own case, his interaction with the president created a disqualifying conflict of interest. The appearance of duplicity was more than a sufficient basis to recuse himself from consideration as special counsel. He should never, in good conscience and under ethical rules, have accepted the position.

It is hard to believe that Mueller had the audacity to accept the position as special counsel that involves, in part, the president's firing of Comey when Mueller was directly involved in the aftermath of that firing. Ask yourself whether it is fair or ethical for a person who was passed over for a job to suddenly reverse course the next day to investigate the man who declined to give him the job. Or perhaps that was Mueller's plan all along. Maybe he had an ulterior motive. Did he devise a sub rosa plan to gather evidence from Trump that he could then use as special counsel, while pretending to be interviewing for the job of FBI director? It is a distinct possibility. If so, it was deceptive and dishonest. Was Comey involved in the plan? Former federal prosecutor Sidney Powell wrote a book about corruption in the Justice Department, *Licensed to Lie.*[37] She surmised that Mueller, Comey, and Rosenstein (who previously worked for Mueller) "communicated extensively and hatched this plan for Mueller to be special counsel—if not prior to Comey's termination, then immediately after it."[38]

The point here is not that this is likely what happened. The point is that, at every step, this investigation was steered by a very small group of people who have known each other and worked together for years. There is nothing remotely approaching independence here.

Compounding the conflict is the evidence that Comey himself may have committed crimes, as outlined in previous chapters. Will Mueller be tempted to ignore the evidence against his friend and choose not to investigate and prosecute Comey? This is another flagrant conflict of interest that compelled Mueller's departure from the case.

Byron York, chief political correspondent for the *Washington Examiner,* who argued that "Mueller should resign" given his conflict of interest, pointed out the irony explained to him by one of several lawyers he consulted on Capitol Hill:

> The whole purpose of the special counsel is to have a prosecutor from outside the government and outside the normal chain of command because inherent conflicts render the Justice Department incapable of handling it. So, now the special counsel is a close friend (mentor/mentee relationship) with the star witness, who by his own admission leaked the memos at least in part to engineer the appointment of a special counsel. Only in Washington. You can't make this stuff up.[39]

Even scrupulously honest people can be influenced in ways they do not recognize themselves. It is the human condition. This is why there are legal and ethical rules that demand recusal based on prior personal relationships. Mueller's conflict is manifest. It is not fair to Trump as the reported subject of the investigation, and it is certainly not fair to the American public. They deserve a legal process that is devoid of partiality or, at the very least, the appearance of it.

By reputation, Mueller is an honest man who has served the public in high positions of government. His service and heroism in the Marine Corps during the Vietnam War is admirable. While there is little doubt about his integrity, it is fair to question his *judgment*. His refusal to recognize two disqualifying conflicts of interest suggest he may harbor a bias. His investigation involving the president of the United States is a serious matter and, therefore, necessitates a special counsel who is beyond reproach in both conduct and appearance. His decision to accept the position in the face of ethical rules discouraging it indicates his judgment has been compromised.

The special counsel regulations require that a person "outside" the Department of Justice, which includes the FBI, be chosen.[40] The reason is quite simple: to avoid conflicts of interest with individuals who work for those departments so that the investigation can be fair and objective. However, in selecting Mueller, Rosenstein picked the ultimate "insider." Mueller spent years as a top official at the Justice Department and also served as director of the FBI. He knew just about everyone and was particularly close to Comey and Rosenstein. While it is true that Mueller was no longer employed at DOJ when he was named special counsel, his various conflicts of interest demonstrate that he was exactly the kind of person the regulations sought to avoid. Many other qualified lawyers would have been better suited for the job and could have brought a greater sense of impartiality and legitimacy to the process.

The unfortunate decision was made, of course, by Rosenstein, who was beset by his conflicts of interest.

ROSENSTEIN SHOULD HAVE RECUSED HIMSELF

Deputy Attorney General Rod Rosenstein has a conflict of interest so acute that no sincere debate could be waged on whether he should have stepped aside. No credible argument can be made in his favor. His refusal to recuse himself in the face of ethical rules that demand his disqualification underscores just how tainted and corrupt the special counsel investigation has become.

It is well established that Mueller is examining whether Trump's firing of Comey was a "corrupt" act intended to obstruct any investigation into Trump-Russian "collusion."[41] The president has insisted that he relied in whole or in part on Rosenstein's recommendation that Comey be terminated for the reasons detailed in the deputy AG's memorandum of May 9, 2017.[42] As Mueller seeks to determine Trump's intent, Rosenstein is a central witness. He had conversations with the president and Attorney General Jeff Sessions about discharging Comey on May 8, 2017, and, perhaps, on other occasions.[43]

Under the regulations (28 C.F.R. 600.7), Rosenstein is in charge of the special counsel's investigation and is Mueller's direct supervisor.[44] As acting attorney general on the case, Rosenstein oversees it, authorizing the scope of the special counsel's probe. He also wields the ultimate power to decide whether a prosecution will be brought, as stated in the regulations:

> The Attorney General may request that the Special Counsel provide an explanation for any investigative or prosecutorial step, and may after review conclude that the action is so inappropriate or unwarranted under established Departmental practices that it should not be pursued.[45]

The *Wall Street Journal* has reported that Mueller's office interviewed Rosenstein in June or July 2017.[46] Think about what that represents. Mueller answers to Rosenstein, but the special counsel and/or his deputies interviewed their boss as a witness. That boss, in turn, plays an instrumental role in deciding whether any case will be leveled against the president based, in whole or in part, on his own testimony.

In an appearance before members of the House of Representatives in May 2017, Rosenstein was questioned as to "who had asked him, if anyone had asked him, to write his memorandum." His answer was quite simple, "That is Bob Mueller's purview."[47] By refusing to disclose who requested that he compose the memo, Rosenstein was admitting that the very investigation he was supervising would involve himself as a witness.

It is fundamentally wrong and unethical for Rosenstein to supervise a case in which he is a chief witness and to simultaneously judge whether a prosecution should be brought. That is the equivalent of a prosecutor putting his boss or himself on the witness stand to testify in the very case he is prosecuting. A lawyer cannot be an investigator, witness, prosecutor, and judge, all rolled into one. Rosenstein plays each of these roles in a single case. His failure to disqualify himself is a serious lapse in judgment. Such conduct is strictly forbidden by the Rules of Professional Conduct.[48]

Understandably, Trump complained about Rosenstein's conflict of interest in a tweet: "I am being investigated for firing the FBI Director by the man who told me to fire the FBI Director. Witch Hunt!"[49] His point was a valid one. In one short sentence, he enunciated the flagrant conflict that prompted Professor Turley to conclude, "If Mueller is investigating Trump for obstruction, Rosenstein should immediately recuse himself."[50] His failure to do so, wrote Turley, "is a glaring ethical omission."[51] The same regulations that required Mueller to step down apply to Rosenstein, as well. Yet, both men refused to do so. This renders the entire special counsel's investigation inherently suspect amid persistent questions of self-interest, political bias, and a lack of objectivity.

Jack Goldsmith, professor of law at Harvard University and a former assistant attorney general, observed that Rosenstein is a "fact witness in any obstruction inquiry" because he "possesses information about the president's beliefs and motives" in firing Comey. He added, "I cannot fathom how, in this light, he remains the supervisor in charge of that investigation, since a reasonable person would question his impartiality."[52]

Rosenstein has been mum on the subject of recusal, except to say that he would do so if needed.[53] The public is supposed to trust him, even in the face of the *appearance* of impropriety due to his conflict as a witness. Rosenstein is a vital witness in the case as evidenced by the report that he was interviewed by the special counsel about Trump's stated motives in firing Comey. As noted in the last chapter, the acting attorney general told investigators that the president knew the firing of the FBI director would not end the Trump-Russia probe, which would discredit any claim that Trump tried to obstruct the case.[54] Rosenstein cannot claim he is merely a peripheral witness who need not be recused. This represents all the more reason why he should have disqualified himself from the case.

Beyond his conflict of interest, Rosenstein had no legal basis to appoint a special counsel to begin with since there was no discernable crime. Joseph diGenova, a former independent counsel and former U.S. Attorney, was unsparing in his condemnation of Rosenstein's actions:

> His conduct from the beginning has been a disgrace. Rod Rosenstein has single-handedly taken away from the sitting President of the United States sixteen months (and counting) of his presidency by his incompetent and fearful conduct. Rosenstein appointed Mueller because he didn't want to make the tough decisions that you'd have to make if you were supervising a case being run by a U.S. Attorney.[55]

But in addition to appointing a special counsel without a legal basis, Rosenstein also approved the team of partisans put together by Mueller to assist him.

MUELLER ASSEMBLED A TEAM OF PARTISANS

Mueller sabotaged his own investigation. He has only himself to blame for ruining what could have been viewed as a credible probe.

He deliberately assembled a team of partisans with a history of political bias who appear determined to undo the results of the 2016 presidential election and drive President Trump from office. Mueller did more than choose to stack the deck of his special counsel staff with biased crusaders. He hired many people with direct, personal connections to Hillary Clinton, transforming what was supposed to be an impartial investigation into an illegitimate and seemingly corrupt one. Rosenstein, who had direct oversight and supervision, condoned the staff selection. He could have objected and demanded that Mueller hire a more balanced team of lawyers and investigators who might have brought a measure of neutrality and objectivity to the investigation. He did not do this.

The media has always been quick to point out that Mueller is a Republican. However, this may or may not be true now. While Mueller was known to be a member of the GOP when he was appointed as FBI director in 2001, "his current party registration is not clear."[56] He has served in government positions under appointment by leaders of both parties. President Bill Clinton nominated Mueller to be U.S. Attorney for the Northern District of California in 1998.[57] In 2011, President Barack Obama asked Mueller to stay on as director of the FBI.[58] His political predilections are something he rarely discloses. But his actions as special counsel showed clear partisanship.

Among the sixteen publicly confirmed lawyers hired by Mueller, not a single one of them is registered as a Republican, according to an in-depth analysis of records by the *Daily Caller*.[59] A similar investigation by Fox News confirmed the same findings.[60] Thirteen of the attorneys are registered Democrats, while three have no party affiliation. The majority of them have donated to Democrats, including several who gave money to the Obama and Clinton presidential campaigns. Only one lawyer, James Quarles, made donations to Republicans, but the two contributions he made pale in comparison to the nearly $38,000 he gave to Democrats.[61] Even more disturbing than the political imbalance on Mueller's staff are the instances of bias that many on the team harbor.

The bias of Peter Strzok and Lisa Page has been well documented herein. They were removed from the team as a consequence.

After them, the most notorious is Andrew Weissmann. After Acting Attorney General Sally Yates, an Obama holdover, defied a direct order from President Trump to defend his newly issued travel ban, Weissmann sent her a message, "I am so proud . . . and in awe. Thank you so much. All my deepest respects."[62] This fawning email to Yates spoke volumes about Weissmann's contempt for Trump and his lack of partiality in his role investigating the president. But that's not all. When Clinton threw what she thought would be an election night victory party at the Javits Center in New York City, Weissmann attended the bash that quickly turned sour when Trump won.[63]

Weissmann is a prosecutor who is known for what has been described as abusive tactics, using the law as a weapon in a sometimes unprincipled quest to convict. He has been accused of suppressing evidence and threatening witnesses.[64] Innocent people have been victimized by his maneuvers. Some of his biggest cases were reversed by higher courts.

As a federal prosecutor, Weissmann helped drive the accounting firm Arthur Andersen out of business, costing tens of thousands of people their jobs, only to have the convictions reversed unanimously by the U.S. Supreme Court. The legal smackdown by the high court underscores what a wrongful prosecution he pursued. But Weissman was undeterred. He went after Merrill Lynch executives, putting them behind bars and destroying their lives, only to have that case also reversed—this time by the Fifth Circuit Court of Appeals.[65]

All of this should have been enough for Weissmann to be banished from the Department of Justice. But Mueller came to the rescue by hiring him as general counsel at the FBI. Weissmann eventually made his way back to the Justice Department as head of the Criminal Fraud Division in the Obama administration. When Mueller became special counsel in the Trump-Russia investigation, he immediately turned to his old "pit bull" of a prosecutor, Weissmann, who joined the team. As former federal prosecutor Powell observed:

> We all lose from Weissmann's involvement. First, the truth plays no role in Weissmann's quest. Second, respect for the

> rule of law, simple decency and following the facts do not appear in Weissmann's playbook. Third, and most important, all Americans lose whenever our judicial system becomes a weapon to reward political friends and punish political foes.[66]

Mueller knew full well what he was doing when he brought Weissmann on board his team of partisans. Weissmann was the "partisan-in-chief."

The special counsel also brought to his staff Jeannie Rhee, who was a partner with Mueller at the law firm WilmerHale. The hiring of Rhee was especially brazen since she defended the Clinton Foundation in a civil racketeering case and donated $5,400 to Clinton's presidential campaign. Thus, Rhee is in a position to bring a prosecution against the person who defeated the candidate she supported and defended. It was a shamelessly bold hire by Mueller, given how obvious a conflict of interest Rhee has. She also worked for Andrew McCabe, who was fired by the FBI for lack of candor. She represented ex-Obama aide Ben Rhodes during the congressional investigation into the Benghazi attack, an inquiry that focused primarily on Clinton's role.[67]

The Clinton connections extend to others on Mueller's team of partisans. Aaron Zebley is on the special counsel team of lawyers, having previously served as Mueller's chief of staff at the FBI. He represented Justin Cooper, who worked for Clinton and set up her private email server, registering the domain in his own name. He also admitted destroying Clinton's BlackBerry devices, "breaking them in half or hitting them with a hammer."[68] He did not have security clearance for any of the classified documents on Clinton's server, yet had access to it. Zebley's involvement in the Clinton case creates yet another appearance of impropriety in the Mueller probe.

Another of Mueller's lawyers is Kyle Freeny, who worked at the Justice Department. She was one of the prosecutors that U.S. District Court Judge Andrew Hanen excoriated in a 2014 case for serious misconduct in her defense of President Obama's executive actions on immigration. In a blistering critique, Freeny and other lawyers were accused of misleading the court in what the judge described as "intentional, serious and material" misconduct.[69] After an apology from the DOJ, Judge Hanen said he would accept the department's claims that the representations to the court were unintentional. Campaign finance records show Freeny donated money to both the Obama and Clinton campaigns.[70]

Just about every member of Mueller's carefully chosen team has ties or allegiances to Democrats, Obama, or the Clintons. In an appearance before the House Judiciary Committee in December 2017, Rosenstein tried in vain to deflect repeated criticism that the special counsel's investigation was tainted by insider bias. When confronted with the evidence that Mueller had hired no Republicans but nearly all Democrats, many of whom donated money to Trump's opponent, Rosenstein insisted that it did not create even the appearance of impropriety which Justice Department regulations prohibit.[71] It was an absurd defense of Mueller's decision to pack his staff with political partisans.

Cheating in a game of cards can involve "stacking the deck"—arranging the cards in a way that advantages yourself while ensuring your opponent loses. It is almost certain that this is the way special counsel Robert Mueller approached his investigation. He chose to hire for his staff a group of lawyers who are Democrats and others, like Strzok and Page, who vented in their messages their hostility toward Trump. Factor into the equation Mueller and Rosenstein's own disqualifying conflicts of interest, and you have an investigation that bears no resemblance to fairness.

The genuine fear is that this special counsel team is so rife with bias that it is more interested in convictions and the political damage that can be wrought on the president than in seeing that justice is done.

EPILOGUE

SUBVERTING THE RULE OF LAW

A government of laws, and not of men.

—John Adams, "Novanglus Papers," No. 7 (1774)

I am not a Trump apologist or sycophant. He will be the first to tell you that.

In the handful of conversations and meetings I had with the president over the past year, he gently reminded me of my earlier reporting that he regarded as negative or critical of him. He had a right to do so, and it was fair. Why, then, would I write a book that is, in large part, a defense of Trump?

In truth, this book is a defense of the rule of law. It came under sustained attack by high government officials who abused their positions of power to subvert our system of justice and undermine the democratic process. Trump was their target and their victim.

As facts emerged, I became incensed that top figures at the FBI and Department of Justice misconstrued the law in a manner that could only have been deliberate. They absolved Hillary Clinton of the felony statutes she so flagrantly violated. They weaponized other laws and regulations to investigate Trump without legal justification in an effort to destroy him.

These officials breached their duty to uphold the law. They compromised essential principles and damaged the nation's trust. Their unconscionable actions inflicted significant harm to the reputation of federal law enforcement and the integrity of the thousands of individuals who serve faithfully and honorably.

The law is inviolate. If it is to function properly as the foundation of democracy, it must be fairly and equitably administered without favor regardless of politics and circumstances. This did not happen. When the legal system is corrupted, all Americans suffer. We lose confidence in our government of laws. As Adams warned, it becomes a government of men who empower themselves to transgress the very rights and liberties we cherish.

In this endeavor, the press plays a vital role. It is an indispensable check on excessive government authority. This is what the Framers envisioned when they embraced the First Amendment in our Bill of Rights. Sadly, it did not always occur in the reporting of this story by the mainstream media. All too often, journalists were complicit in advancing or excusing the abuse of power. They convicted Trump in the court of public opinion without evidence cognizable in law. This is another reason why I decided to write this book.

Fueled by the FBI's unwarranted investigation, reporters became advocates. They condemned Trump for the least infraction and inflated any encounter with persons connected, however tangentially, to Russia. Many in the media, motivated by their own bias, drove the constant narrative that Trump and those in his orbit "colluded" with Moscow to steal the 2016 presidential election from Clinton. When the new president fired FBI director James Comey, the action was immediately interpreted and portrayed by the press as irrefutable proof of obstruction of justice.

Some journalists accused Trump of being "Putin's puppet" or "a de facto agent" of the Kremlin.[1] Articles were published implying that Trump was a "Manchurian candidate."[2] The unfairness of the media was palpable. Rank speculation was treated as fact. Stories were agenda-driven, not information-driven. Indictment and impeachment were persistent themes in both print and television news. At every turn, the media demonized the president by declaring him guilty of the conjured crime of "collusion." It became their favored truncheon.

Examples of this abound. When Trump's lawyer, Jay Sekulow, was interviewed on *ABC News* by anchor George Stephanopoulos, the following exchange took place about a Trump adviser who spoke to a person with Russian contacts:

> **SEKULOW:** There is no crime of "collusion." What is a violation of law here?
>
> **STEPHANOPOULOS:** Collusion is cooperation![3]

Viewers were left with the impression that the act of talking with someone constituted "collusion" which equaled the commission of a crime. Yet, there was no evidence of a secret agreement to achieve some fraudulent or illegal or deceitful purpose. It was all too easy for the anchor to insinuate a crime by using a word that connotes a crime. The *appearance* of criminality seemed sufficient for many reporters.

Others took to the airwaves to imply or pronounce criminal

"collusion" without ever identifying what law might have been broken or the specific facts that merited such a legal conclusion:

> **CARL BERNSTEIN (ON CNN):** I think this is a potentially more dangerous situation than Watergate. We're at a dangerous moment. And that's because we are looking at the possibility that the president of the United States and those around him during an election campaign colluded with a hostile foreign power to undermine the basis of our democracy—free elections.[4]

> **DAN RATHER (ON MSNBC):** Donald Trump is afraid. He's trying to exude power and strength. He's afraid of something that Mueller and the prosecutors are going to find out. A political hurricane is out there at sea for him. We'll call it Hurricane Vladimir, if you will, the whole Russian thing. It is approaching Category Four.[5]

> **JAKE TAPPER (CNN ANCHOR):** "This is evidence of willingness to commit collusion. That's what this is on its face."[6]

> **LAWRENCE O'DONNELL (MSNBC HOST):** "Donald Trump now sits at the threshold of impeachment."[7]

These were, by no means, the only instances in which the media proclaimed Trump and his associates culpable of criminal "collusion." One anchor averred with certainty, "they're *confessing* to colluding with the Russians."[8] Another speculated that "people might go to jail for the rest of their lives."[9] Still another insisted that "they're acting super guilty because they're guilty."[10] But guilty of what specifically? That vexing tidbit of information was conveniently omitted.

Throughout the Trump presidency, the media obsessed over "collusion" without defining it. They assumed it was a crime that surely must exist buried somewhere in the vast body of dusty law books. One anchor became so giddy over the prospect of Trump's arrest, she fanaticized on air about the day he would barricade himself inside the White House as federal marshals banged on the door to take him into custody.[11] The president's imminent demise became daily fodder.

There has been no shortage of media malpractice in the age of Trump. Motivated by political bias and personal animus, some journalists were relentless in their quest to prove the president's illegitimacy and drive him from office. They abandoned objectivity and suspended their sense of fairness. They allowed enmity to cloud their judgment. In the process, the media squandered credibility, its only currency. It is no wonder that many Americans have little trust in journalists to be honest in their reporting.[12]

The people who should read this book probably won't. Democrats, the liberal media, and the legion of Trump-haters have convinced themselves that his election was misbegotten. But they are intellectually dishonest in believing that the president must have committed some crime in connection with Russia. Most have never bothered to examine the facts or consult a statute. Their confidence in "collusion" is bereft of proof. They accept it as a matter of faith driven by their own bias and teased by hope out of ignorance.

I have no illusion that what I have written in these pages will persuade them otherwise. The anti-Trump crowd is so adamant in their disdain that no amount of reason will reach them. Yet, they are equally blind to the astonishing level of corruption by those in government who sought to sabotage the president with false claims and drive him from office. Individuals who are sworn to uphold the law manipulated it for their own partisan purpose. Their self-righteousness compulsion to correct what they perceived to be wrong in the election of Trump serves as the quintessential example of the arrogance of power.

Several months have passed since *The Russia Hoax* went to print in early June of 2018. Since then, significant new evidence has emerged that only corroborates the thesis of this book. The inspector general at the Department of Justice, Michael Horowitz, uncovered more incriminating text messages between the FBI's Peter Strzok and Lisa Page who played leading roles in both the Clinton and Trump investigations. These texts demonstrated a virulent pro-Clinton and anti-Trump bias that led the I-G to conclude they harbored a "willingness to take official action to impact" the 2016 election and stop Trump from becoming president.[13] Comey's actions were also roundly condemned by the I-G for his misjudgments, bias, insubordination, and unprofessionalism.[14] The investigation of Comey continues, as of this writing.

Comey knew the FBI was incorruptible. This is precisely why he seized control of the investigation of the Clinton case instead of allowing the FBI field office to conduct the probe. By commandeering control of the investigation, Comey and his confederates could twist the facts and contort the law to dictate the outcome they desired. Clinton was the beneficiary of what FBI Deputy Director Andrew McCabe reportedly described as the "HQ special."[15] McCabe, of course, was eventually fired and is the subject of a criminal probe.

Peter Strzok was also fired by the FBI after his egregious misconduct was revealed by the I-G. His lover and colleague Lisa Page resigned. Other Comey lieutenants who were involved in the Clinton and Trump cases also resigned after they were demoted or before they could be terminated.

After Page departed the FBI, she was questioned in a closed-door deposition before two congressional committees. Those who attended confirmed that Page, the FBI's lead lawyer in the Russia case, readily admitted that bureau investigators had discovered no evidence of collusion between Russia and the Trump campaign during the entire time that the FBI handled the case.[16] Thus, when Robert Mueller was appointed, there was no evidence of a crime, which is a legal prerequisite for invoking the special counsel. For this reason, his investigation was illegitimate under federal regulations.

Inexplicably, there is one person who has not, as yet, lost his job. That is, Associate Deputy Attorney General Bruce Ohr, who was instrumental to the scheme to frame Trump. After *The Russia Hoax* was published, Ohr's machinations came into greater focus when he testified in a private session before the House Oversight and Judiciary Committees. It was learned that Ohr bypassed his superiors at the Justice Department and worked covertly with Christopher Steele to exploit the anti-Trump "dossier." Sixty-three pages of his texts, emails, and handwritten notes show extensive contacts between Ohr and Steele both before and after the 2016 election.[17]

Even after Steele was fired by the FBI for lying about his contacts with reporters, Ohr maintained intensive communications with him. The former British spy kept feeding his bogus information while simultaneously fretting to Ohr that his actions might be "exposed."[18] It was a direct violation of FBI procedures to use a discredited source who had also broken an agreement with the FBI. To circumvent its rules, the bureau secretly used Ohr as a conduit to continue its relationship with Steele.

Ohr would communicate with Steele, then endeavor to pass the information along to Comey's FBI. Strzok confirmed in his congressional testimony that "the FBI received documents and material from Mr. Ohr."[19] Amazingly, Ohr was peddling the false Steele document while benefiting from it financially. How? His wife, Nellie Ohr, worked on the "dossier" as an employee of Fusion GPS and was getting paid to do so by the Clinton campaign. Money was going into the Ohrs' bank account at the same time Bruce Ohr was using his position at the Justice Department to advance this false intelligence document. When called to testify before Congress, Nellie Ohr invoked the spousal privilege to avoid answering many of the integral questions about the "dossier" and how it was surreptitiously fed to the FBI and DOJ.[20] Glenn Simpson, the co-founder of Fusion GPS, invoked his Fifth Amendment right against self-incrimination and refused to answer any questions at all.[21]

As noted earlier in the book, Bruce Ohr well knew that Steele was a nefarious and unreliable character. Records show that in September of 2016 Steele told Ohr that he was "desperate that Donald Trump not get elected and was passionate about him not being president."[22] Given such an acute bias and motivation to lie, all contact with Steele should have been terminated and his "dossier" tossed in the garbage. However, both Ohr and the FBI were determined to exploit Steele and his unverified documents to damage Trump.

Significantly, Steele was not only receiving money from the Clinton campaign, but he was also on the payroll of the FBI from January 1, 2016 to November 1, 2016. Records obtained by Judicial Watch through a lawsuit reveal that he was paid eleven times when he met with the FBI.[23] Hence, the FBI and Clinton's campaign were working in concert to use a discredited source, who composed a fictitious document based on supposed Russian information to harm her political opponent in the 2016 election. It is confounding that Clinton and her campaign have never been investigated for violating federal election laws and other relevant criminal statutes. If anyone was "colluding," coordinating, or conspiring with Russia, it was Clinton, the FBI, and the DOJ.

Documents also show that Steele kept feeding the FBI uncorroborated allegations about Trump—even after Trump became president—all the way through May 15, 2017.[24] Why? It is reasonable to conclude that the FBI was desperate to verify the unverified "dossier." This phony material from Steele served as the basis for launching its Trump-Russia investigation and the foundation of its warrant to wiretap a Trump campaign associate. Having violated FBI regulations by using unverified information in a warrant application and thereby committing a fraud on the FISA court, the FBI was surely hoping to cure its wrongdoing belatedly. It's called covering your tracks.

Under Comey's authority, the FBI declared that Steele was "not suitable for use as a Confidential Human Source (CHS)," and he was fired by the bureau in the Fall of 2016.[25] Yet, two years later, when Comey was selling his book (and his moral purity) on a nationwide tour, he vouched for Steele calling him "credible."[26] Soon thereafter, documents surfaced that belied Comey's claim.

Under pressure from Congress, FISA records were finally released in the summer of 2018. They showed what many have long suspected—Steele's dubious "dossier" was "a major component of the 2016 surveillance warrant" to spy on a Trump campaign associate.[27] Without verifying the veracity of its contents as demanded by federal regulations, the FBI simply declared that Steele was "credible" and "reliable." Clearly, he was not since he'd already been sacked for lying and was deemed unsuitable as a source. The

FBI also informed the FISA court that Steele was not the direct source for a Yahoo story that was cited in its warrant application. This declaration was made to the judges not just once, but on four separate occasions. British court records prove that the FBI's affirmation, repeatedly made, was utterly untrue.[28]

There is little doubt that the court was deceived when it issued a warrant to spy on Carter Page. There was never so much as a hearing.[29] The court accepted as truthful the sworn representations by the FBI and DOJ that the information and statements presented were honest, accurate, and corroborated. They were not.

Lisa Page was not the only former FBI official to implicate others at the agency who engaged in wrongdoing during the Trump investigation. James Baker, who was general counsel at the bureau and worked closely with Comey, resigned on the same day as Page. Months later in closed-door testimony, he delivered what was described as "explosive" information.[30] Two lawmakers who attended the session told Fox News that Baker admitted the Russia case was handled in an "abnormal fashion" reflecting "political bias."[31] He also confirmed the FBI surely knew that Clinton and Democrats were behind the unsubstantiated "dossier" because they gave the bureau both documents and a thumb drive that were intended to damage Trump.[32] The partisan source of the information was never disclosed to the FISA court when the FBI obtained permission to spy on a Trump campaign associate.

Under oath, Baker also reportedly implicated Deputy Attorney General Rod Rosenstein in a plot to secretly record conversations with President Trump and recruit cabinet members to invoke the Twenty-fifth Amendment in a concerted effort to depose the president in the spring of 2017, shortly after Trump fired Comey based on Rosenstein's memo recommending the termination.[33] The story, first reported by the *New York Times*, stated that Baker thought

Rosenstein "was serious about taping Trump."[34] Baker based his conclusion on information given to him directly by both McCabe and Page who were privy to the alleged plot.[35] Rosenstein denied it all, but then suggested that his words were taken out of context and were merely "sarcastic."[36] As of this writing, the deputy attorney general has resisted all efforts by Congress to have him testify under oath about the subject.

Proposing to secretly record the president is, at the very least, a violation of regulations that govern a security clearance at the White House. More important, Rosenstein's suspected actions further taint the special counsel investigation of Trump that the deputy attorney general was then overseeing. How can anyone view the probe as fair, objective, and neutral? Mueller's investigation has been compromised and contaminated by Rosenstein's behavior, the special counsel's refusal to disqualify himself in the face of blatant conflicts of interest, and the team of partisans assembled to pursue Trump.

As all of the chicanery and schemes unraveled, it became increasingly apparent that saboteurs at the FBI and Justice Department had worked furiously to undue the election results and frame Trump for "colluding" with Russia to win the presidency. They conjured a false case based on a fabricated "dossier" paid for by Hillary Clinton's campaign and composed by an ex-British spy who was fired for lying. They misappropriated that document to launch the Trump-Russia investigation without probable cause. They then exploited the same "dossier" to wiretap a Trump campaign associate, Carter Page. In the process, they concealed essential evidence and deceived FISA judges, perpetrating a fraud on the court.

By claiming the mantle of "classified" status, Rosenstein, the FBI, and the DOJ have repeatedly refused to produce records that Congress lawfully subpoenaed. In what is surely a collective effort to hide or cover up evidence of their own wrongdoing, they have obstructed Congress and obscured the truth. If made public, these documents would likely shed valuable light on the pervasive bias that led to Clinton being cleared, the paucity of evidence in launching the Trump investigation, the deceptions employed to spy on Carter Page, and exculpatory evidence that would have quashed the Russia probe.

A sane, decent, and civilized democracy must never operate completely in the furtive shadows of concealment by its own government. The Constitution grants Congress the "implied powers" to hold the executive branch officials accountable for their actions by lifting the veil of secrecy. This is the only way to prevent malfeasance and deceit.

Rosenstein, of course, has every reason to suppress the records sought. They may well incriminate him. He affixed his signature to the final renewal of the FISA warrant application to continue spying on Page, likely without sufficient evidence. He vouched for the authenticity and truthfulness of the information contained therein. But if it was largely based on a "dossier" that was unverified and uncorroborated, Rosenstein was complicit in misusing his authority in pursuit of a president who was unjustifiably and unlawfully targeted by the FBI and DOJ. As explained in Chapter 11, he appointed a special counsel without proper authority. He then refused to recuse himself as a key witness in the very case over which he presides. All along the way, he has been thwarting Congress by withholding records they are entitled to have.

The day after the midterm elections in early November of 2018, Attorney General Jeff Sessions, at long last turned in his resignation after it was justifiably requested by President Trump. By wrongfully recusing himself from the Russia case, Sessions caused immeasurable damage to the Trump presidency and to the nation.

He became the personification of misfeasance or nonfeasance. His actions, or lack thereof, were born of incompetence. Sessions rarely exhibited the kind of leadership skills that are demanded of the nation's leading law enforcement official.

More often than not, Sessions was missing in action. As President Trump quite accurately remarked in an interview, "I don't have an attorney general."[37] This is the reason Sessions was finally, if belatedly, shown the door. America and the president of the United States both deserve to have a functioning Justice Department and a competent attorney general.

Sessions refused to reopen the Clinton email case and present compelling evidence of possible criminality by her to a federal grand jury. When evidence emerged that top officials at the FBI abused their power to launch an investigation of Trump-Russia "collusion" without credible evidence as federal regulations require, Sessions failed to take action. When he learned of FISA abuse, he did absolutely nothing about it. Sessions committed multiple mistakes that led to the appointment of Special Counsel Robert Mueller. He also endorsed the effort to obstruct Congress. Sessions's departure will allow a new attorney general to repair the considerable damage he left in his wake. The next A-G must end the persistent cover-ups and work assiduously to uphold the rule of law, while holding accountable those within the FBI and DOJ who appear to have violated it.

It has been both challenging and frustrating to compose a book as events are still unfolding. It is akin to the struggle of keeping pace with a fast-moving train. There will be more revelations to come, as congress and the Justice Department examines the venal conduct of those who cleared Clinton and targeted Trump. It may take months, if not years, before a true accounting is divulged. This is how cover-ups work. The truth is slow to emerge. Until then, the story told here presents more than sufficient facts for the reader to reach the inexorable conclusion that Trump was a victim, not the villain.

There was never any plausible evidence that Trump or his campaign collaborated with Russia to win the presidency. The FBI had no legal basis to initiate its investigation. Facts were invented or exaggerated. Laws were perverted or ignored. The law enforcers became law breakers. Comey's scheme to trigger the appointment of his friend as special counsel was a devious maneuver by an unscrupulous man. His insinuation that the president obstructed justice was another canard designed to inflame the liberal media. Sure enough, they became witting accessories.

When powerful forces in government abuse their positions of trust to subvert the legal process, and when the media acts in concert to condone or conceal corrupt behavior, democracy is threatened. Reverence to the rule of law is lost.

This is the story of *The Russia Hoax*.

GREGG JARRETT

NOVEMBER 2018

ACKNOWLEDGMENTS

Winston Churchill once said that composing a book is an adventure. It begins as a toy, then evolves into an amusement, a mistress, a master, and, eventually, a tyrant. He was right. It is not an experience for the faint of heart.

There are many people to whom I am grateful. David Limbaugh, a fine lawyer and author in his own right, was kind enough to provide counsel and advice in negotiating the contract to write this book. I am indebted to him for his generosity.

Eric Nelson, executive editor at HarperCollins, was instrumental in shaping the content into a more readable book. He displayed both wisdom and patience for a first-time author. He also had the courage to tell me that not all of the roughly one hundred thousand words I submitted were good enough for print. His edits were sensible, yet judicious. I am thankful for his guidance. A special thanks to the rest of the team at HarperCollins.

Mike Plante furnished research and valuable assistance in many of the interviews that were conducted. I am most appreciative to all of the former federal prosecutors and FBI officials who gave freely of their time to be interviewed. Their insights and candor improved the book immeasurably. A special thank you to Joseph diGenova, Victoria Toensing, and Doug Burns.

My sincere thanks to Sean Hannity, who liked the notion of a book from the outset. He was a constant source of support, providing his friendship and good ideas as the chapters developed. More than anyone else, he unraveled and reported many of the corrupt acts identified in these pages. I owe him a great deal. His staff, Porter Berry, Tiffany Fazio, Bret Zoeller, Christen Limbaugh, Alyssa Moni, Francesca Nestande, and Lynda McLaughlin, were especially encouraging.

Lou Dobbs, a driving force at the Fox Business Network, contributed his enthusiasm for the book and his endorsement of me personally as a guest on his program. His great staff of Anne McCarton, Lilah Sabalones, Abigail Penn, Travis Altman, Robert Regan, Michael Biondi, Mike LaMarca, and Sabrina Lee have been exceedingly kind.

Many others at Fox News deserve recognition. My friend Lynn Jordal Martin, senior opinion editor for the Fox News website, kept urging me to write a book. Refet Kaplan, Greg Wilson, Janet Cawley, and Morgan Debelle Duplan gave prominence to my opinion columns. Kimberly Sialiano helped with the computer format and other advice. Courtney Stein Lesskis supplied needed legal research in a pinch. Mary Kate Cribbin lent a friendly ear to my frustrations. Rich Reichmuth tolerated my anxieties and kept coaxing me to move forward. Lis Wiehl, a veteran author and good friend, offered wisdom when needed, which was fairly often.

My wife and two daughters were an endless supply of love, encouragement, and support. Without them, I would never have embarked on this "adventure." My sister, Janet, kept faith in me, as she has since I can remember. My mother's spirit is always a presence in me.

Finally, many of the quotations at the beginning of each chapter came from the books my late father left me. Joseph W. Jarrett bequeathed to me his vast law library, which has dwindled over the years. But I retained a few special books to which I turned when in need of quotes to help frame the issues at hand. My father also gave me his love of the law and its navigating principles. He was the finest lawyer I ever knew.

On a more personal note, I recognize my failings plainly and regret them. Like many people, I have faltered in life. Still, I chose to write this book myself. These are my words within the pages. If mistakes were made, they are mine alone.

NOTES

CHAPTER 1: HILLARY CLINTON'S EMAIL SERVER

1. Meghan Keneally, Liz Kreutz, and Shushannah Walshe, "Hillary Clinton Email Mystery Man: What We Know About Eric Hoteham," ABC News, March 5, 2015; Josh Gerstein, "The Mystery Man Behind Hillary's Email Controversy," *Politico*, March 4, 2015.
2. Department of State, Foreign Affairs Manual, 5 FAM 8710 ("Users must not load classified information or Sensitive but Unclassified (SBU) information onto unclassified systems . . .").
3. Congressional Testimony of James Comey, House Committee on Oversight and Government Reform, July 7, 2016 (Question: "Did Hillary Clinton give noncleared people access to classified information?" Comey: "Yes."); Richard Lardner, Associated Press, "Witnesses Refuse to Testify in Hearing on Clinton's Email," September 13, 2016.
4. Report by Federal Bureau of Investigation, "Clinton Email Investigation"—11 and 12, September 2, 2016; Report by Office of Inspector General, Department of State, "Office of the Secretary: Evaluation of Email Records Management and Cybersecurity Requirements," May 25, 2016.
5. Senator Hillary Clinton's Remarks to the Newspaper Association of America's Annual Conference in Washington, D.C., April 15, 2008 ("When I am president I will empower the federal government to operate from a presumption of openness, not secrecy. We will adopt a presumption of openness and Freedom of Information Act requests and urge agencies to release information quickly if disclosure will do no harm"), available at http://web.archive.org/web/20080904230344/htpp://www.fas.org/sgp/news/2008/04/clinton041508.html.
6. Report by Federal Bureau of Investigation, "Clinton Email Investigation"—9 and 10, September 2, 2016.
7. Interview of President Barack Obama, CBS News, aired March 8, 2015; Reena Flores, "Obama Weighs In on Hillary Clinton's Private Emails," CBS News,

March 7, 2015; Kyle Cheney, "Clinton Aide Talked of Needing to 'Clean' Up Obama's Comments on Email Server," *Politico*, October 25, 2016; Lisa Lerer, "Top Obama Aides Knew About Clinton's Private Email in 2009," Associated Press, June 30, 2015.

8. Evelyn Rupert, "Obama Used Pseudonym in Emails with Clinton," *The Hill*, September 23, 2016; Josh Gerstein and Nolan D. McCaskill, "Obama Used a Pseudonym in Emails with Clinton, FBI Documents Reveal," *Politico*, September 23, 2016; Daniel Halper, "Obama Used Pseudonym to Talk with Hillary on Private Server," September 24, 2016.
9. Ibid.
10. Chuck Ross, "Top Clinton Aides Face No Charges After Making False Statements to FBI," *Daily Caller*, December 4, 2017.
11. Report by Office of Inspector General, Department of State, "Office of the Secretary: Evaluation of Email Records Management and Cybersecurity Requirements," May 25, 2016; Steven Lee Myers and Eric Lichtblau, "Hillary Clinton Is Criticized for Private Emails in State Department Review," *New York Times*, May 25, 2016.
12. Report by Federal Bureau of Investigation, "Clinton Email Investigation," p. 12 ("State Diplomatic Security Service [DS] instructed Clinton that because her office was in a SCIF [Sensitive Compartmented Information Facility], the use of mobile devices in her office was prohibited. Interviews of three former DS agents revealed Clinton stored her personal BlackBerry in a desk in DS 'Post 1,' which was located within the SCIF on Mahogany Row. State personnel were not authorized to bring their mobile devices into Post 1, as it was located within the SCIF.").
13. Robert O'Harrow Jr., "How Clinton's Email Scandal Took Root," *Washington Post*, March 27, 2016.
14. Ibid.
15. Ibid. Select Committee on Benghazi, Final Report, July 8, 2016, available at benghazi.House.gov/NewInfo.
16. Rachael Bade, "Meet the Clinton Insider Who Screened Hillary's Emails," *Politico*, September 4, 2015.
17. Statement by FBI director James B. Comey on the Investigation of Secretary Hillary Clinton's Use of a Personal E-Mail System, FBI National Press Office, July 5, 2016.
18. Michael S. Schmidt, "Hillary Clinton Used Personal Email Account at State Department, Possibly Breaking Rules," *New York Times*, March 2, 2015.
19. Zeke J. Miller, "Transcript: Everything Hillary Clinton Said on the Email Controversy," *Time*, March 10, 2015.
20. Ibid.
21. Eugene Scott, "Hillary Clinton on Emails: 'The Facts Are Pretty Clear,'" CNN, July 28, 2015.
22. Statement by FBI director James B. Comey on the Investigation of Secretary Hillary Clinton's Use of a Personal E-Mail System, FBI National Press Office, July 5, 2016.

23. Josh Feldman, "Hillary: Use of Personal Email 'Clearly Wasn't the Best Choice,'" *Mediaite,* August 26, 2015; Glenn Kessler, "Clinton's Claims About Receiving or Sending 'Classified Material' on Her Private Email System," *Washington Post,* August 27, 2015.
24. Glenn Kessler, "Clinton's Claims."
25. Testimony of James B. Comey, FBI Director, House Committee on Oversight and Government Reform, July 7, 2016.
26. Harper Neidig, "Clinton to FBI: Didn't Know Parenthetical 'C' Stood for Confidential," *The Hill,* September 2, 2016.
27. Interview with Bill Gavin, former assistant director of the FBI, February 18, 2018.
28. Interview with Oliver "Buck" Revell, former FBI assistant deputy director, March 19, 2018.
29. Hillary Clinton interview with CNN, July 7, 2015, available at http://cnnpressroom.blogs.cnn.com/2015/07/07/cnn-exclusive-hillary-clintons-first-national-interview-of-2016-race/.
30. Glenn Kessler, "Hillary Clinton's Claim That 'Everything I Did on Emails Was Permitted,'" *Washington Post,* July 9, 2015.
31. Report by Office of Inspector General, Department of State, "Office of the Secretary: Evaluation of Email Records Management and Cybersecurity Requirements," May 25, 2016, available at https://oig.state.gov/system/files/esp-16-03.pdf; Ken Dilanian, "Clinton Broke Federal Rules with Email Server, Audit Finds," NBC News, May 25, 2016.
32. Ibid. Report by Federal Bureau of Investigation, "Clinton Email Investigation," September 2, 2016; Rosalind S. Helderman and Tom Hamburger, "State Department Inspector General Report Sharply Criticizes Clinton's Email Practices," *Washington Post,* May 25, 2016.
33. Report by Office of Inspector General, Department of State, "Office of the Secretary: Evaluation of Email Records Management and Cybersecurity Requirements," May 25, 2016, available at https://oig.state.gov/system/files/esp-16-03.pdf.
34. Michael S. Schmidt and Matt Apuzzo, "Hillary Clinton Emails Said to Contain Classified Data," *New York Times,* July 24, 2015.
35. Jason Donner, "Judge Orders State Department to Work on Recovering Emails, Suggests Clinton Violated Policy," Fox News, August 20, 2015; Josh Gerstein, "Judge Says Hillary Clinton's Private Emails Violated Policy," *Politico,* August 20, 2015; Michael S. Schmidt, "Judge Says Hillary Clinton Didn't Follow Government Email Policies," *New York Times,* August 20, 2015.
36. Pamela K. Browne and Catherine Herridge, "Hillary Clinton Signed Non-Disclosure Agreement to Protect Classified Information While Secretary of State," Fox News, November 7, 2015; Brendan Bordelon, "Clinton Acknowledged Penalties for 'Negligent Handling' of Classified Information in State Department Contract," *National Review,* November 6, 2015.
37. Chuck Ross, "Document Completely Undermines Hillary's Classified Email Defense," *Daily Caller,* November 6, 2015; Jeryl Bier, "Hillary Signed She Received

Briefing on Classified Information, but Told FBI She Hadn't," *Weekly Standard,* September 2, 2016.

38. http://www.foxnews.com/politics/interactive/2015/03/05/state-department-cable-june-28-2011.html.
39. Report by Federal Bureau of Investigation, "Clinton Email Investigation," p. 18, September 2, 2016; Byron York, "From FBI Fragments, A Question: Did Team Clinton Destroy Evidence Under Subpoena," *Washington Examiner,* September 3, 2016; DeTroy Murdock, "Obstruction of Justice Haunts Hillary's Future," *National Review,* September 8, 2016.
40. Report by Federal Bureau of Investigation, "Clinton Email Investigation," p. 19, September 2, 2016.
41. Cause of Action, "Legal Analysis of Former Secretary of State Hillary Clinton's Use of a Private Server to Store Email Records," August 24, 2015, available at http://causeofaction.org/wp-content/uploads/2015/08/Hillary-Clinton-Email-Memo-8.24.15.pdf.
42. Interview with Steve Pomerantz, former assistant director of the FBI, March 14, 2018.
43. Interview with Joseph diGenova, former U.S. attorney for the District of Columbia and former independent counsel, January 26, 2018.
44. United States Code, 44 U.S.C. 3101 et al., "Records Management by Agency Heads, General Duties"; Foreign Affairs Manual, 5 F.A.M. 441(h)(2), et al.
45. United States Code, 18 U.S.C. 641, "Public Money, Property or Records."
46. United States Code, 18 U.S.C. 2071(b), "Concealment, Removal or Mutilation Generally."
47. Memorandum of Patrick F. Kennedy, Under Secretary of State for Management, "Senior Officials' Records Management Responsibilities," August 28, 2014 (Kennedy: "All records generated by Senior Officials belong to the Department of State"), available at http://goo.gl/bT6qLz; U.S. Department of State, Department Notice, Message from Under Secretary for Management Patrick F. Kennedy Regarding State Department Records Responsibilities and Policy, October 17, 2014, available at http://goo.gl/sDixrN.
48. *United States v. Rosen,* 352 Fed. Su—915 (1972).
49. David E. Kendall, letter to Trey Gowdy, chairman of the House Select Committee on Benghazi, March 27, 2015, available at http://goo.gl/jXpS5x; Lauren French, "Gowdy: Clinton Wiped Her Server Clean," *Politico*, March 27, 2015.
50. United States Code, 18 U.S.C. 1001, "Statements or Entries Generally."
51. United States Code, 18 U.S.C. 1505, "Obstruction of Justice Proceedings Before Departments, Agencies and Committees"; United States Code, 18 U.S.C. 1515(b), "Definitions for Certain Provisions, General Provision."
52. Statement by FBI Director James B. Comey on the Investigation of Secretary Hillary Clinton's Use of a Personal E-Mail System, FBI National Press Office, p. 3, July 5, 2016.
53. United States Code, 18 U.S.C. 793(f), "Gathering, Transmitting or Losing Defense Information."

54. Ibid., Sections (d) and (e).
55. United States Code, 18 U.S.C. 1924(a), "Unauthorized Retention of Removal of Classified Documents or Material."
56. United States Code, 18 U.S.C. 371, "Conspiracy to Commit Offense or to Defraud United States"; United States Code, 18 U.S.C. 286, "Conspiracy to Defraud the Government with Respect to Claims."
57. Zeke J. Miller, "Transcript: Everything Hillary Clinton Said on the Email Controversy," *Time,* March 10, 2015; Alexandra Jaffe and Dan Mercer, "Hillary Clinton: I Used One Email 'For Convenience,'" CNN, March 11, 2015; Amy Chozick and Michael S. Schmidt, "Hillary Clinton Tries to Quell Controversy Over Private Email," *New York Times*, March 10, 2015.
58. Ibid.
59. United States Code, 44 U.S. C. 3101, "Records Management by Agency Heads, General Duties"; Code of Federal Regulations, 36, CFR 1236.22; Department of State, Foreign Affairs Manual, 5 FAM 443.1.
60. United States Code, 18 U.S.C. 2071, "Concealment, Removal, or Mutilation Generally."
61. United States Code, 18 U.S.C. 1924, "Unauthorized Removal and Retention of Classified Documents or Material"; United States Code, 18 U.S.C. 798, "Disclosure of Classified Information"; United States Code, 18 U.S.C. 793(f), "Gathering, Transmitting or Losing Defense information."
62. Byron Tau and Peter Nicholas, "State Department Lacked Top Watchdog During Hillary Clinton Tenure," *Wall Street Journal,* March 24, 2015.
63. Interview with Joseph diGenova, former U.S. attorney for the District of Columbia and former independent counsel, January 26, 2018.
64. Interview with Doug Burns, former assistant U.S. attorney for the Eastern District of New York, March 23, 2018.

CHAPTER 2: COMEY CONTORTS THE LAW TO CLEAR CLINTON

1. Statement by FBI Director James B. Comey on the Investigation of Secretary Hillary Clinton's Use of a Personal Email System, FBI National Press Office, July 5, 2016.
2. Ibid.
3. Ibid.
4. Interview with Danny Coulson, former deputy assistant director of the FBI, February 12, 2018.
5. Interview with Steve Pomerantz, former assistant director of the FBI, March 14, 2018.
6. Carrie Johnson, "Officials Scrutinized Over Classified Information, but Rarely Found Criminal," National Public Radio, April 7, 2016.
7. Ibid.
8. Ibid.
9. Associated Press, "Navy Engineer Sentenced for Mishandling Classified Material," *Navy Times,* July 29, 2015.
10. Jennifer Rizzo, "Feds Torpedo Navy Sailor's 'Clinton Defense,'" CNN, August 17, 2016.

11. Kristina Wong, "Troops Using 'Clinton Defense' in Classified Information Cases," *The Hill,* August 20, 2016.
12. John Bowden, "Trump Pardons Navy Sailor Who Used 'Clinton Defense,'" *The Hill,* March 9, 2018.
13. Kelley Beaucar Vlahos, "Ex-Officials Prosecuted for Mishandling Government Information See 'Double Standard' in Clinton Case," Fox News, August 17, 2015.
14. Statement by FBI Director James B. Comey on the Investigation of Secretary Hillary Clinton's Use of a Personal Email System, FBI National Press Office, July 5, 2016.
15. Ann E. Maimow, "Ex-NSA Contractor Pleads Not Guilty to Spying Charges in Federal Court," *Washington Post*, February 14, 2017.
16. Statement by FBI Director James B. Comey on the Investigation of Secretary Hillary Clinton's Use of a Personal Email System, FBI National Press Office, July 5, 2016.
17. Interview with Danny Coulson, former deputy assistant director of the FBI, February 12, 2018.
18. Ibid.
19. United States Code, 18 U.S.C. 793(f), "Gathering, Transmitting or Losing Defense Information."
20. Some lawyers and legal commentators have argued that prosecutors must still prove "intent" under the "gross negligence" provision of the above statute. Think about that for just a moment. How exactly does a person *intend* to do something in a *grossly negligent* manner? People either intend to commit an act or they perform an act in a grossly negligent manner. Can both really be accomplished contemporaneously? They cannot. The former reflects a specific state of mind, while the latter tends to reflect the absence of it.

    Those who conflate "intent" with "gross negligence" have relied mistakenly on a 1941 Supreme Court case called *Gorin v. United States*. In *Gorin*, the high court was interpreting the explicit language of the *original* Espionage Act of 1917 when it held that "intent" must be proven. However, seven years after the *Gorin* decision, the Act was amended and supplanted by several sections of 18 U.S.C. 793, including the "gross negligence" provision. Thus, *Gorin* has no application or relevance to section (f) because that section of the law did not exist at the time the court rendered its ruling.

    In the decades since the *Gorin* case, courts have explained that "intent" is required for other sections of the statute, but not section (f), which established a lower threshold of "gross negligence." For example, in *United States v. McGuiness*, the court stated that "it is clear that Congress intended to create a hierarchy of offenses against national security, ranging from 'classic spying' to mere losing classified materials through gross negligence."

    In the famous "Pentagon Papers" case, Justice Byron White wrote that the district court erred in relying on *Gorin* when it determined that the government needed to prove "only willful and knowing conduct." White noted that *Gorin* "arose under other parts of the predecessor (law) . . . parts that imposed different intent standards not repeated in" the successor statute.

Sources: Dan Abrams, "Trump Is Wrong, Hillary Clinton Shouldn't Be Charged Based On What We Know Now," Lawandcrime.com, March 2, 2016 (originally published on LawNewz.com, January 29, 2016); Rachel Stockman, "Despite 'Bombshell' New Emails, Hillary Clinton Probably Still Didn't Commit a Crime," Lawandcrime.com, October 29, 2016. *United States v. McGuinness,* 35 M.J. 149, 153 (CMA 1992). *New York Times Co. v. United States,* 403 U.S. 713 (1971).

21. United States Code, 18 U.S. C. 793(d) and (e), "Gathering, Transmitting or Losing Defense Information."
22. United States Code, 18 U.S. C. 793, sections (a) through (d).
23. Stephen L. Vladeck, "Prosecuting Leaks Under U.S. Law," Justsecurity.org, November 2015.
24. Statement by FBI Director James B. Comey on the Investigation of Secretary Hillary Clinton's Use of a Personal Email System, FBI National Press Office, July 5, 2016.
25. Testimony of James B. Comey, Director of the FBI, before the House Oversight Committee, July 7, 2016, available at https://www.youtube.com/watch?v=xQ6hJb9SNlw.
26. Ibid.
27. United States Code, 18 U.S.C. 1924(a), "Unauthorized Removal and Retention of Classified Documents or Material."
28. Zeke J. Miller, "Transcript: Everything Hillary Clinton Said on The Email Controversy," *Time,* March 10, 2015.
29. Report by Federal Bureau of Investigation, "Clinton Email Investigation," September 2, 2016; Report by Office of Inspector General, Department of State, "Office of the Secretary: Evaluation of Email Records Management and Cybersecurity Requirements," May 25, 2016.
30. Pamela K. Brown and Catherine Herridge, "Hillary Clinton Signed Non-Disclosure Agreement to Protect Classified Information While Secretary of State," Fox News, November 7, 2015; Brendan Bordelon, "Clinton Acknowledges Penalties for 'Negligent Handling' of Classified Information in State Department Contract," *National Review,* November 6, 2015.
31. Andrew McCarthy, "Restoring the Rule of Law to the Protection of Classified Information," *National Review,* January 6, 2018.
32. Sonam Sheth, "The FBI Agent Mueller Ousted Was Behind 2 Critical Turning Points in the Clinton and Trump-Russia Investigations," *Business Insider,* December 4, 2017.
33. John Solomon, "Early Comey Draft Accused Clinton of Gross Negligence on Emails," *The Hill,* November 6, 2017.
34. Jake Gibson and Justin Berger, "Comey Edits Revealed: Remarks on Clinton Probe Were Watered Down, Documents Show," Fox News, December 14, 2017; Max Kutner, "Comey's FBI Initially Believed Clinton Was 'Grossly Negligent' in Handling Classified Emails," *Newsweek,* November 7, 2017.
35. Interview with Doug Burns, former assistant United States attorney for the Eastern District of New York, March 23, 2018.

36. Senate Judiciary Committee letter to FBI Director Christopher Wray, August 30, 2017, available at https://www.judiciary.senate.gov/imo/media/doc/2017-08-30%20CEG%20+%20LG%20to%20FBI%20(Comey%20Statement).pdf.
37. Interview with Fred Tecce, former assistant U.S. attorney in Philadelphia and special assistant U.S. attorney at the Department of Justice, March 28, 2018.
38. Ibid.
39. United States Code, 18 U.S.C. 1505, "Obstruction of Proceedings Before Departments, Agencies, and Committees."
40. United States Code, 18 U.S.C. 1515(b), "Definitions for Certain Provisions; General Provisions."
41. Hearing Before the Committee on the Judiciary, "Oversight of the Federal Bureau of Investigation," testimony of James B. Comey, September 28, 2016, available at https://judiciary.house.gov/wp-content/uploads/2016/09/114-91_22125.pdf.
42. United States Code, 18 U.S.C. 1621, "Perjury Generally;" United States Code, 18 U.S.C. 1001, "Statements or Entries Generally."
43. Interview with Bill Gavin, former assistant director of the FBI, February 18, 2018.
44. Interview with Oliver "Buck" Revell, former FBI assistant deputy director, March 19, 2018.
45. Christopher Sign, "U.S. Attorney General Loretta Lynch, Bill Clinton Meeting Privately in Phoenix Before Benghazi Report," KNXV-TV, ABC15, June 29, 2016 (updated July 4, 2016).
46. Matt Zapotosky, "Attorney General Declines to Provide Any Details on Clinton Email Investigation," *Washington Post*, July 12, 2017.
47. Code of Federal Regulations, 28 C.F.R. 45.2, "Disqualification Arising from Personal or Political Relationship."
48. Douglas Ernst, "FBI Ordered 'No Cellphones' to Airport Witnesses of Lynch-Clinton Meeting, Reporter Says," *Washington Times*, July 1, 2016.
49. Jordan Sekulow, "DOJ Document Dump to ACLJ on Clinton Lynch Meeting: Comey, FBI Lied, Media Collusion, Spin and Illegality," American Center for Law and Justice, August, 2017, available at https://aclj.org/government-corruption/doj-document-dump-to-aclj-on-clinton-lynch-meeting-comey-fbi-lied-media-collusion-spin-and-illegality.
50. The Editors, "When Loretta Met Bill," *Weekly Standard*, August 11, 2017.
51. Zachary Cohen, "Comey Cites Lynch-Clinton Meeting for Lost Faith in Justice Investigation," CNN, May 3, 2017.
52. Testimony of FBI Director James Comey, Senate Judiciary Committee, May 3, 2017, available at https://www.washingtonpost.com/news/post-politics/wp/2017/05/03/read-the-full-testimony-of-fbi-director-james-comey-in-which-he-discusses-clinton-email-investigation/?utm_term=.8a81b5b77e70.
53. Peter Baker, "Comey Raises Concerns About Loretta Lynch's Independence," *New York Times*, June 8, 2017; Tal Kopan, "Comey: Lynch Asked for Clinton Investigation to Be Called a Matter," CNN, June 8, 2017; Ed O'Keefe, "Comey Repeats That Lynch Asked Him to Describe Clinton Investigations as a 'Matter,'"

Washington Post, June 8, 2017; Kelly Cohen, "James Comey: Loretta Lynch Told Me Not to Call Clinton Email Probe an 'Investigation,'" *Washington Examiner*, June 8, 2017.

54. Ibid.
55. Interview with Joseph diGenova, former U.S. attorney for the District of Columbia and former independent counsel, January 26, 2018.
56. Malia Zimmerman and Adam Housley, "FBI, DOJ Roiled by Comey, Lynch Decision to Let Clinton Slide by on Emails, Says Insider," Fox News, October 13, 2016.
57. Ibid.
58. Ibid.
59. Ibid.
60. United States Code, 18 U.S.C. 207, "Restrictions on Former Officers, Employees, and Elected Officials of the Executive and Legislative Branches."
61. Dan McLaughlin, "Why Did Hillary Clinton Spend Three and a Half Hours This Morning With FBI?," *National Review*, July 2, 2016; Dan Berman, Evan Perez, and Pamela Browne, "Clinton Questioned by FBI as Part of Email Probe," CNN, July 3, 2016.
62. Tal Kopan and Evan Perez, "FBI Releases Hillary Clinton Email Report," CNN, September 2, 2016; Aaron Blake, "Hillary Clinton Told the FBI She Couldn't Recall Something More Than Three Dozen Times," *Washington Post*, September 2, 2016.
63. Malia Zimmerman and Adam Housley, "FBI, DOJ Roiled by Comey, Lynch Decision to Let Clinton Slide by on Emails, Says Insider," Fox News, October 13, 2016.
64. Paul Sperry, "FBI Agents Are Ready to Revolt Over the Cozy Clinton Probe," *New York Post,* October 6, 2016.
65. Ibid.
66. Ibid.

CHAPTER 3: "THE FIX"

1. FBI National Press Office, "Andrew McCabe Named Deputy Director of the FBI," January 29, 2016, available at https://www.fbi.gov/news/pressrel/press-releases/andrew-mccabe-named-deputy-director-of-the-fbi.
2. Lauretta Brown, "Judicial Watch Releases New DOJ Documents Showing McCabe's Conflict of Interest in Clinton Investigation," Townhall, November 22, 2017.
3. Fox News, Catherine Herridge contributing, "Clinton Ally Helped Fund Campaign of Key FBI Official's Wife," Fox News, October 24, 2016; Judicial Watch, "New Documents Show FBI Deputy Director McCabe Did Not Recuse Himself from the Clinton Email Scandal Investigation Until Week Before Presidential Election," November 3, 2017, available at https://www.judicialwatch.org/press-room/press-releases/judicial-watch-new-documents-show-fbi-deputy-director-mccabe-not-recuse-clinton-email-scandal-investigation-week-presidential-election/.

4. Letter from Senator Charles Grassley, chairman of Senate Judiciary Committee, to James B. Comey, Jr., director of Federal Bureau of Investigation, March 28, 2017, available at https://www.grassley.senate.gov/news/news-releases/grassley-examines-potential-conflicts-top-fbi-official%E2%80%99s-role-russia-collusion.
5. Lauretta Brown, "Judicial Watch Releases New DOJ Documents Showing McCabe's Conflict of Interest in Clinton Investigation," Townhall, November 22, 2017.
6. Devlin Barrett, "Clinton Ally Aided Campaign of FBI Official's Wife," *Wall Street Journal*, October 24, 2016.
7. Kevin Johnson, "FBI Documents: Andrew McCabe Had No Conflict in Hillary Clinton Email Probe," *USA Today*, January 5, 2018.
8. Natasha Bertrand, "Emails Released by the FBI Shed New Light on Deputy Director's Recusal from Clinton Probe," *Business Insider*, January 9, 2018.
9. Ibid.
10. Lauretta Brown, "Judicial Watch Releases New DOJ Documents Showing McCabe's Conflict of Interest in Clinton Investigation," Townhall, November 22, 2017; Judicial Watch, "New Documents Show FBI Deputy Director McCabe Did Not Recuse Himself from the Clinton Email Scandal Investigation Until Week Before Presidential Election," November 3, 2017, available at https://www.judicialwatch.org/press-room/press-releases/judicial-watch-new-documents-show-fbi-deputy-director-mccabe-not-recuse-clinton-email-scandal-investigation-week-presidential-election/.
11. Judicial Watch, "New Documents Show FBI Deputy Director McCabe Did Not Recuse Himself from the Clinton Email Scandal Investigation Until Week Before Presidential Election."
12. Ibid.
13. Circa.com, "The Face of FBI Politics: Bureau Boss McCabe Under Hatch Act Investigation," June 27, 2017, available at https://www.circa.com/story/2017/06/27/the-face-of-fbi-politics-bureau-boss-mccabe-under-hatch-act-investigation; letter by Charles E. Grassley, chairman of Committee on the Judiciary to Rod Rosenstein, deputy attorney general, December 1, 2017, available at https://www.judiciary.senate.gov/imo/media/doc/2017-12-01%20CEG%20to%20DOJ%20(McCabe%20Hatch%20Act%20Emails).pdf.
14. Letter from Senator Charles Grassley, chairman of Senate Judiciary Committee, to James B. Comey, Jr., director of Federal Bureau of Investigation, March 28, 2017, available at https://www.grassley.senate.gov/news/news-releases/grassley-examines-potential-conflicts-top-fbi-official%E2%80%99s-role-russia-collusion.
15. John Solomon, "Early Comey Draft Accused Clinton of Gross Negligence on Emails," *The Hill*, November 6, 2017.
16. Manu Raju and Jeremy Herb, "Top FBI Official Grilled on Comey, Clinton in Hill Testimony," CNN, December 22, 2017.
17. Kelly Cohen, "FBI Deputy Director Andrew McCabe to Retire: Report," *Washington Examiner*, December 23, 2017.

18. Manu Raju and Jeremy Herb, "Top FBI Official Grilled on Comey, Clinton in Hill Testimony," CNN, December 22, 2017.
19. Barnini Chakraborty, "FBI's McCabe Faces GOP Calls for Ouster, Ahead of Closed-Door Testimony," Fox News, December 19, 2017.
20. Steven T. Dennis and Billy House, "FBI's McCabe Meets with House Panel as Grassley Calls for His Firing," Bloomberg, December 18, 2017; Chuck Ross, "FBI Deputy Director 'Ought to Be Replaced,' Says Top Senate Republican," *Daily Caller*, December 19, 2017.
21. Devlin Barrett and Karoun Demirjian, "Facing Republican Attacks, FBI's Deputy Director Plans to Retire Early Next Year," *Washington Post*, December 23, 2017; Adam Goldman, "Andrew McCabe, FBI's Embattled Deputy, Is Expected to Retire," *New York Times*, December 23, 2017.
22. Judson Berger and Jake Gibson, "FBI Deputy Director Andrew McCabe 'Removed' From the Bureau," Fox News, January 29, 2018.
23. Reuters Staff, "Statement by Attorney General on Firing of FBI's McCabe," Reuters, March 17, 2018.
24. Ibid.
25. Alex Pappas, "IG Report Faults Ex-FBI Official McCabe for Lying Four Separate Times, Congressman Says," Fox News, March 30, 2018.
26. United States Code, 18 U.S.C. 1621, "Perjury, Generally"; United States Code, 18 U.S.C. 1001, "Statements or Entries Generally."
27. CNN Staff, "Read: Former FBI Deputy Director Andrew McCabe's Statement on His Firing," CNN, March 17, 2018.
28. *Washington Post* Staff, "Read the Full Transcript of FBI Director James Comey In Which He Discusses Clinton Email Investigation," *Washington Post*, May 3, 2017.
29. Interview with James Kallstrom, former assistant director of the FBI, Sunday Morning Futures with Maria Bartiromo, March 18, 2018, available at https://www.youtube.com/watch?v=TiRryItvVRo; Many Mayfield, "Former Assistant FBI Director: High-Ranking People Throughout the Government Had a Plot to Protect Hillary Clinton From Being Indicted," *Washington Examiner*, March 18, 2018; Tim Hains, "Kallstrom: Brennan and Obama Are the People Who Committed Felonies, Not General Flynn," RealClearPolitics, March 18, 2018.
30. Interview with Oliver "Buck" Revell, former FBI assistant deputy director, March 19, 2018.
31. John Solomon, "Early Comey Draft Accused Clinton of Gross Negligence on Emails," *The Hill*, November 6, 2017.
32. Karoun Demirjian and Devlin Barrett, "Top FBI Official Assigned to Mueller's Russia Probe Said to Have Been Removed After Sending Anti-Trump Texts," *Washington Post*, December 2, 2017; Michael S. Schmidt, Matt Apuzzo, and Adam Goldman, "Mueller Removed Top Agent in Russia Inquiry Over Possible Anti-Trump Texts," *New York Times*, December 2, 2017; Jake Gibson, "'Over 10,000 Texts' Between Ex-Muller Team Officials Found, After Discovery of Anti-Trump Messages," Fox News, December 6, 2017; Brooke Singman, Alex

Pappas, Jake Gibson, "More Than 50,000 Texts Exchanged Between FBI Officials Strzok and Page, Sessions Says," Fox News, January 22, 2018.
33. FoxNews.com, "Ex-Mueller Aides' Texts Revealed: Read Them Here," Fox News, December 13, 2017.
34. Ibid.
35. Laura Jarrett and Evan Perez, "FBI Agent Dismissed from Mueller Probe Changed Comey's Description of Clinton to 'Extremely Careless,'" CNN, December 4, 2017.
36. Jake Gibson and Justin Berger, "Comey Edits Revealed: Remarks on Clinton Probe Were Watered Down, Documents Show," Fox News, December 14, 2017.
37. Kelly Cohen, "DOJ Provides Congress with Hundreds of Texts Between Ex-Mueller Team Agent Peter Strzok and Alleged Mistress Lisa Page," *Washington Examiner*, January 11, 2018.
38. FoxNews.com, "Ex-Mueller Aides' Texts Revealed: Read Them Here," Fox News, December 13, 2017.
39. Samuel Chamberlain, "Newly Released Texts Between Ex-Mueller Team Members Suggest They Knew Outcome of Clinton Email Probe in Advance," Fox News, January 21, 2018.
40. Brooke Singman, Alex Pappas, and Jake Gibson, "More Than 50,000 Texts Exchanged Between FBI Officials Strzok and Page, Sessions Says," Fox News, January 22, 2018.
41. Statement by FBI Director James B. Comey on the Investigation of Secretary Hillary Clinton's Use of a Personal Email System, FBI National Press Office, July 5, 2016.
42. Matt Zapotosky, "Attorney General Pledges to Accept FBI and Justice Findings in Clinton Email Probe," *Washington Post*, July 1, 2016.
43. Julian Hattem, "Attorney General Says She Will Defer to FBI on Clinton Emails," *The Hill*, July 1, 2016.
44. Ibid.
45. John Solomon, "Early Comey Draft Accused Clinton of Gross Negligence on Emails," *The Hill*, November 6, 2017.
46. Brooke Singman, "New Texts Show 'Fix Was In' for Clinton Email Probe, GOP Lawmakers Say," Fox News, January 26, 2018.
47. Jake Gibson, "FBI Lovers' Latest Text Messages: Obama 'Wants to Know Everything,'" Fox News, February 7, 2018.
48. Fox News Staff, "Exclusive: President Barack Obama on 'Fox News Sunday,'" Fox News, April 10, 2016.
49. Jake Gibson, "FBI Lovers' Latest Text Messages: Obama 'Wants to Know Everything,'" Fox News, February 7, 2018.
50. Ibid.
51. Bre Payton, "What We Know About the Dismissed FBI Agent Who Interviewed Michael Flynn," *Federalist,* December 5, 2017.
52. Chuck Ross, "Top Clinton Aides Face No Charges After Making False Statements to FBI," *Daily Caller*, December 4, 2017.
53. Ibid.

54. United States Code, 18 U.S.C. 1001, "Statements or Entries Generally."
55. Chuck Ross, "Top Clinton Aides Face No Charges After Making False Statements to FBI," *Daily Caller*, December 4, 2017.
56. Sonam Sheth, "The FBI Agent Mueller Ousted Was Behind 2 Critical Turning Points in the Clinton and Trump Russia Investigations," *Business Insider*, December 4, 2017.
57. Editorial Board, "What the Strzok?," *New York Post*, December 6, 2017.
58. United States Code, 18 U.S.C. 1505, "Obstruction of Proceedings Before Departments, Agencies, and Committees."
59. United States Code, 5 U.S.C. 7323, "Political Activity Authorized; Prohibitions."
60. Statement of House Judiciary Committee Chairman Bob Goodlatte at Oversight Hearing with Deputy Attorney General Rod Rosenstein, December 13, 2017, available at https://goodlatte.house.gov/news/documentsingle.aspx?DocumentID=1064.

CHAPTER 4: CLINTON GREED AND "URANIUM ONE"

1. Liz Kreutz, "Hillary Clinton Defends High-Dollar Speaking Fees," ABC News, June 9, 2014, available at http://abcnews.go.com/Politics/hillary-clinton-defends-high-dollar-speaking-fees/story?id=24052962.
2. Dan Alexander, "How the Clintons Have Made $230 Million Since Leaving the White House," *Forbes,* October 13, 2015.
3. Ibid.
4. Philip Rucker, Tom Hamburger, and Alexander Becker, "How the Clintons Went From 'Dead Broke' to Rich: Bill Earned $104.9 Million for Speeches," *Washington Post*, June 26, 2014.
5. Bob Fredericks, "How Foreign Cash Made Bill and Hillary 'Filthy Rich,'" *New York Post*, April 20, 2015.
6. Peter Schweizer, *Clinton Cash: The Untold Story of How and Why Foreign Governments and Businesses Helped Make Bill and Hillary Rich* (New York: Harper, 2015), p. 15.
7. Ibid., p. 17.
8. CQ Transcripts Wire, "Transcript: Hillary Clinton Confirmation Hearing, *Washington Post*, January 13, 2009.
9. Jonathan Swan, "Seven Ways the Clinton Foundation Failed to Meet Its Transparency Promises," *The Hill*, August 27, 2016.
10. Jonathan Allen, "Exclusive: Despite Hillary Clinton Promise, Charity Did Not Disclose Donors," Reuters, March 19, 2015.
11. Ibid., p. 182.
12. Victor Davis Hanson, "How the Clintons Got Rich Selling Influence While Decrying Greed," *National Review*, July 26, 2016.
13. Bruce Pannier, "Kazakhstan Long Term President to Run in Show Election—Again," *Guardian,* March 11, 2015.
14. "Fox News Reporting: The Tangled Clinton Web," Fox News Channel, aired April 24, 2015.
15. Schweizer, p. 29.

16. Ibid., pp. 49 and 54.
17. Ibid., p. 55.
18. Mike Halers, "Russian Company Seeks to Buy U.S. Uranium Mining Operations," CNN, September 25, 2010.
19. Schweizer, pp. 44 and 45.
20. Ibid., pp. 54 and 55.
21. Ibid., p. 55.
22. Ibid., p. 51.
23. Jo Becker and Mike McIntire, "Cash Flowed to the Clinton Foundation Amid Russian Uranium Deal," *New York Times*, April 23, 2015.
24. Ibid.
25. Schweizer, p. 48.
26. Ibid.
27. Ibid.
28. Rosalind S. Helderman and Tom Hamburger, "1,100 Donors to a Canadian Charity Tied to Clinton Foundation Remain Secret," *Washington Post*, April 28, 2015.
29. Callum Borchers, "Making Sense of Russia, Uranium One and Hillary Clinton, as Congress Opens an Investigation," *Washington Post*, October 24, 2017.
30. John Solomon and Alison Spann, "FBI Uncovered Russian Bribery Plot Before Obama Administration Approved Controversial Nuclear Deal with Moscow," October 17, 2017.
31. Ibid.
32. Ibid.
33. John Solomon, "Uranium One Informant Makes Clinton Allegations to Congress," *The Hill*, February 7, 2018.
34. John Solomon, "FBI Informant Gathered Years of Evidence on Russian Push for U.S. Nuclear Fuel Deals, Including Uranium One, Memos Show," *The Hill*, November 20, 2017.
35. John Solomon and Alison Spann, "FBI Informant Says U.S. Had Evidence to Block Billions in Nuclear Deals for Russia," *The Hill*, March 23, 2018.
36. John Solomon and Alison Spann, "Informant Provided FBI Evidence Russia Aided Iran Nuclear Program During Obama Years," *The Hill*, March 26, 2018.
37. John Solomon and Alison Spann, "Russian Uranium Informant Says FBI Sought New Information from Him About the Clintons," *The Hill*, March 22, 2018.
38. John Solomon, "Uranium One Informant Makes Clinton Allegations to Congress."
39. Jessica Kwong, "Russia Routed Millions to Influence Clinton in Uranium Deal, Informant Tells Congress," *Newsweek,* February 8, 2018; David Krayden, "FBI Informant: US Lobbyists Paid by Russia to Influence Clinton on Uranium One," *Daily Caller*, February 8, 2018.
40. Brooke Singman, "Uranium One Informant Says Moscow Paid Millions in Bid to Influence Clinton," Fox News, February 8, 2018.

41. Ibid.
42. Ibid.
43. United States Code, 18 U.S.C. 201(b), "Bribery of Public Officials and Witnesses."
44. United States Code, 18 U.S.C. 201(c), "Bribery of Public Officials and Witnesses."
45. United States Code, 18 U.S.C. 1341, "Frauds and Swindles."
46. United States Code, 18 U.S.C. 1343, "Elements of Wire Fraud."
47. United States Code, 18 U.S.C. 1346, "Definition of Scheme or Artifice to Defraud."
48. United States Code, 18 U.S.C. 1952, "Interstate and Foreign Travel or Transportation in Aid of Racketeering Enterprises."
49. United States Code, 18 U.S.C. 1956, "Laundering of Monetary Instruments."
50. United States Code, 18 U.S.C. 1961–1968, "Racketeer Influenced and Corrupt Organizations."
51. United States Department of Justice, Offices of the United States Attorneys, 109 (RICO Charges), available at https://www.justice.gov/usam/criminal-resource-manual-109-rico-charges.
52. Tim Winter, Pete Williams, and Ken Dilanian, "Prosecutors Ask FBI Agents for Information on Uranium One Deal," NBC News, December 21, 2017.
53. Letter from Jeff Sessions, Office of the Attorney General to Senator Charles Grassley, Representatives Robert Goodlatte and Trey Gowdy, March 29, 2018; Brooke Singman, "Sessions: Federal Prosecutor Evaluating Alleged FBI, DOJ Wrongdoing, No Second Special Counsel for Now," Fox News, March 29, 2018.
54. Stephen Braun and Eileen Sullivan, "Many Donors to Clinton Foundation Met with Her at State," Associated Press, August 24, 2016.
55. Rosalind S. Helderman and Tom Hamburger, "Inside 'Bill Clinton Inc': Hacked Memorandum Reveals Intersection of Charity and Personal Income," *Washington Post*, October 26, 2016; Rosalind S. Helderman, Spencer S. Hsu, and Tom Hamburger, "Emails Reveal How Foundation Donors Got Access to Clinton and Her Close Aides at State Department," *Washington Post*, August 22, 2016.
56. Memorandum by Douglas Band, p. 2, November 16, 2011, available at https://www.politico.com/f/?id=00000158-048f-da25-a55e-efaf30b60000.
57. Ibid., p. 10.
58. Ibid.
59. James V. Grimaldi and Rebecca Ballhaus, "UBS Deal Shows Clinton's Complicated Ties," *Wall Street Journal*, July 30, 2015; Conor Friedersdorf, "Hillary Helps a Bank—and Then It Funnels Millions to the Clintons," *Atlantic*, July 31, 2015.
60. United States Code, 18 U.S.C. 201, "Bribery of Public Officials and Witnesses."
61. Fox News (Cody Derespina contributed), "'Bill Clinton Inc.': Email Details How Top Aides Helped Make Ex-President Rich," Fox News, October 26, 2016.
62. Rebecca Ballhaus, "Newly Released Emails Highlight Clinton Foundation's Ties to State Department," *Wall Street Journal*, August 10, 2016.

63. Katie Bo Williams and Jonathan Easley, "Memo Reveals Interplay Between Clinton Foundation, Personal Business," *The Hill*, October 26, 2016.
64. Fox News, "Clinton Aide Says Foundation Paid for Chelsea's Wedding, WikiLeaks Emails Show," Fox News, November 6, 2016.
65. Editorial, "The Clinton Foundation Is Dead—but the Case Against Hillary Isn't," *Investor's Business Daily,* January 19, 2017.
66. Isabel Vincent and Melissa Klein, "Donations to Clinton Foundation Fell by 37 Percent," *New York Post,* November 20, 2016; Khaleda Rahman, "Donations to the Clinton Foundation Plummeted Amid Hillary's Failed Run for the Presidency," *Daily Mail,* November 20, 2016.
67. Isabel Vincent, "Donations to Clinton Foundation Plunged Along with Hillary's Election Defeat," *New York Post,* November 18, 2017.
68. Michael Sainato, "The Clinton Foundation Shuts Down Clinton Global Initiative," *Observer,* January 15, 2017.

CHAPTER 5: THE FRAUDULENT CASE AGAINST DONALD TRUMP

1. Office of the Director of National Intelligence, "Background to 'Assessing Russian Activities and Intentions in Recent U.S. Elections': The Analytic Process and Cyber Incident Attributions," January 6, 2017, available at https://www.dni.gov/files/documents/ICA_2017_01.pdf.
2. Ibid., p. 2.
3. Ibid., p. 5.
4. Ibid., p. 1.
5. Ibid.
6. Laura Jarrett and Evan Perez, "FBI Agent Dismissed from Mueller Probe Changed Comey's Description of Clinton to 'Extremely Careless,' " CNN, December 4, 2017.
7. Sonam Sheth, "The FBI Agent Mueller Ousted Was Behind 2 Critical Turning Points in the Clinton and Trump-Russia Investigations," December 4, 2017.
8. Byron York, "After Mysterious 'Insurance Policy' Test, Will Justice Department Reveal More on FBI Agent Bounced from Mueller Probe?," *The Hill,* January 13, 2017.
9. Ibid.
10. Ibid.
11. Tim Hains, "Rep. Jim Jordan Grills FBI Director Wray About Peter Strzok: Did He Use Steele Dossier to Obtain FISA Warrant?," RealClearPolitics, December 7, 2017.
12. Jake Gibson, " 'Over 10,000 Texts' Between Ex-Muller Team Officials Found, After Discovery of Anti-Trump Messages," December 6, 2017.
13. Interview with Steve Rogers, former Senior Naval Intelligence Officer for the FBI National Joint Terrorism Task Force in Washington, D.C., February 7, 2018.
14. "The Attorney General's Guidelines for Domestic FBI Operations"; also found at United States Code, 28 U.S.C. 509, 510, 533, and 534; and Executive Order 12333, available at https://www.justice.gov/archive/opa/docs/guidelines.pdf.
15. Federal Bureau of Investigation, "FBI Domestic Investigations and Operations Guide" (DIOG), available at https://vault.fbi.gov/FBI%20Domestic%20Inves

tigations%20and%20Operations%20Guide%20%28DIOG%29/fbi-domestic-investigations-and-operations-guide-diog-2013-version.

16. "The Attorney General's Guidelines for Domestic FBI Operations," page 12; also found at United States Code, 28 U.S.C. 509, 510, 533, and 534; and Executive Order 12333, available at https://www.justice.gov/archive/opa/docs/guidelines.pdf.
17. Ibid., p. 16.
18. Ibid., p. 18.
19. Ibid., p. 22.
20. Matthew Kazin, "No Official Intel Used to Start FBI Probe into Trump Campaign-Russia Collusion: Rep. Nunes," Fox Business Network, April 22, 2018.
21. Andrew McCarthy, "No, the FBI Was Not a Trump Partisan," *National Review*, May 6, 2017.
22. Ibid., p. 13.
23. Andrew McCarthy, "Fighting the Politicized, Evidence-Free 'Collusion with Russia' Narrative," *National Review*, May 24, 2017.
24. *Washington Post* Staff, "Full Transcript: FBI Director James Comey Testifies on Russian Interference in 2016 Election," *Washington Post*, March 20, 2017.
25. Interview with Fred Tecce, former assistant director U.S. attorney in Philadelphia and special assistant U.S. attorney at the Department of Justice, March 28, 2018.
26. Letter from Senator Harry Reid to FBI Director James Comey, August 27, 2016, available at https://archive.org/details/ReidLetterToComey08272016.
27. Paul Sperry, "Exclusive: CIA Ex-Director Brennan's Perjury Peril," RealClearInvestigations, February 11, 2018.
28. Sara A. Carter, "Collusion Delusion: New Documents Show Obama Officials, FBI Coordinated in Anti-Trump Probe," p. 2, March 29, 2018, available at https://saraacarter.com/new-documents-suggest-coordination-by-obama-white-house-cia-and-fbi-in-trump-investigation/.
29. Ibid., p. 2.
30. Ibid., p. 5.
31. Brooke Singman, "Grassley Rips Strzok-Page Redactions Amid Mystery Text: Obama 'White House Is Running This,'" Fox News, May 23, 2018.
32. Josh Rogin, "Democrats Ask the DBI to Investigate Trump Advisers' Russia Ties," *Washington Post*, August 30, 2016.
33. Letter from Senator Harry Reid to FBI Director James Comey, October 30, 2016, available at https://web.archive.org/web/20161031183809/http://www.reid.senate.gov/wp-content/uploads/2016/10/Letter-to-Director-Comey-10_30_2016.pdf.
34. Aaron Blake, "Harry Reid's Incendiary Claim About 'Coordination' Between Donald Trump and Russia," *Washington Post*, October 31, 2016.
35. John Brandt, "Reid Repeats Rumor on Senate Floor That Romey Paid No Taxes, Campaign Denies," Fox News, August 2, 2012.
36. Louis Jacobson, "Harry Reid Says Anonymous Source Told Him Mitt Romney Didn't Pay Taxes for 10 Years," PolitiFact, August 6, 2012; Glenn Kessler,

"4 Pinocchios for Harry Reid's Claim About Mitt Romney's Taxes," *Washington Post* Fact Checker, August 7, 2012.

37. Ashley Condianni, "Harry Reid Doesn't Regret Accusing Mitt Romney of Not Paying Taxes," CNN, March 31, 2015.
38. James B. Comey, "Statement for the Record, Senate Select Committee on Intelligence," p. 2, June 8, 2017, available at https://www.intelligence.senate.gov/sites/default/files/documents/os-jcomey-060817.pdf.
39. Transcript: FBI Director James Comey Testimony Before House Intelligence Committee, March 20, 2017, available at https://www.washingtonpost.com/news/post-politics/wp/2017/03/20/full-transcript-fbi-director-james-comey-testifies-on-russian-interference-in-2016-election/?utm_term=.0db15bbfea4f.
40. James B. Comey, "Statement for the Record, Senate Select Committee on Intelligence," p. 6, June 8, 2017, available at https://www.intelligence.senate.gov/sites/default/files/documents/os-jcomey-060817.pdf.
41. Sherman Antitrust Act (1890), United States Code, 15 U.S.C. 1–7, "Trusts, Etc., in Restrain of Trade Illegal; Penalty."
42. Gregg Jarrett, "What Is Robert Mueller Investigating (Since Collusion Is Not a Crime)?," Fox News, May 23, 2017.
43. Jon Greenberg, "Fox News Host Wrong That No Law Forbids Russia-Trump Collusion," PolitiFact, May 31, 2017.
44. United States Code, 18 U.S.C. 1346, "Definition of 'Scheme or Artifice to Defraud'."
45. *Skilling v. United States,* 561 U.S. 358 (2010).
46. Jon Greenberg, "Fox News Host Wrong That No Law Forbids Russia-Trump Collusion," PolitiFact, May 31, 2017.
47. Ibid.
48. United States Code, 18 U.S.C. 610, "Coercion of Political Activity."
49. United States Code, 18 U.S.C. 1030, "Fraud and Related Activity in Connection With Computers."
50. Elana Schor, "Burr: Russia Probe Will Expose Erroneous Reporting," *Politico*, October 5, 2017.
51. Associated Press, "Burr: No evidence So Far of Trump Campaign-Russia Collusion," Associated Press, May 12, 2017.
52. Interview by Jake Tapper of Senator Mark Warner, CNN, *State of the Union,* June 4, 2017, available at http://www.cnn.com/TRANSCRIPTS/1706/04/sotu.01.html.
53. Associated Press, "The Latest: Nunes Says There's 'No Evidence of Collusion,'" *Washington Post*, February 2, 2018.
54. Nicolas Fandos, "Despite Mueller's Push, House Republicans Declare No Evidence of Collusion," *New York Times*, March 12, 2018.
55. Rowan Scarborough, "Democrats' Case for Trump-Russia Grand Conspiracy Crumbles with Lack of Evidence," *Washington Times,* December 27, 2017.
56. Marisa Schultz, "Schiff: Evidence in Russia Probe Is 'Damning,'" *New York Post,* December 10, 2017; Rebecca Savransky, "Schiff: Evidence of Coordination Between Trump Campaign and Russia 'Pretty Damning,'" *The Hill*, December 10,

2017; Erin Kelly, "Adam Schiff: There Is 'Ample Evidence' of Collusion Between Trump Campaign, Russians," *USA Today,* February 14, 2018.

57. Rachel Stotzfoos, "Watch: Schiff Concedes There Is Still No Proof of Trump-Russia Collusion," *Federalist,* March 1, 2018.
58. Ian Schwartz, "No Evidence of Collusion, So Democrats Have Pivoted to Obstruction of Justice," RealClearPolitics, December 5, 2017.
59. Sam Stein and Ryan Grim, "Harry Reid: The Trump Campaign 'Was In On' Russia's Election Hacking," *Huffington Post,* December 12, 2016.
60. Eli Watkins, "Wasserman Schultz: Trump Aides' Russian Contact 'Reeks of Collusion,'" CNN, February 15, 2017.
61. Chuck Ross, "Maxine Waters 'Guarantees' That Trump Colluded with Russia," *Daily Caller,* September 21, 2017.
62. Jay Yarow, "Hillary Clinton Says 'Americans' Guided Russia's Attack on Her Campaign, Suggesting Trump's Campaign 'Colluded' with Russia," CNBC, June 1, 2017.
63. Peter Hasson, "Still No Evidence of Trump-Russia Collusion as 2017 Draws to a Close," *Daily Caller,* December 29, 2017.
64. *Meet the Press,* "Full Clapper: 'No Evidence' of Collusion Between Trump and Russia," NBC News, March 5, 2017; Ian Schwartz, "Clapper: 'No Direct Evidence of Political Collusion Between the Trump Campaign and the Russians," RealClearPolitics, May 30, 2017.
65. *Washington Post* Staff, "Full Transcript: FBI Director James Comey Testimony on Russian Interference in 2016 Election," *Washington Post,* March 20, 2017.
66. Testimony of Former CIA Director John Brennan Before Congress, CNN Transcripts, May 23, 2017, available at http://edition.cnn.com/TRANSCRIPTS/1705/23/cnr.04.html; Tim Hains, "Gowdy Grills Brennan: Do you Have Evidence of Trump-Russia Collusion or Not? Brennan: 'I Don't Do Evidence,'" RealClearPolitics, May 23, 2017.
67. Ken Dilanian, "Clinton Ally Says Smoke, But No Fire: No Russia-Trump Collusion," NBC News, March 16, 2017.
68. National Travel & Tourism Office, "U.S. Resident Travel Abroad Historical Visitation—Outbound," available at https://travel.trade.gov/outreachpages/download_data_table/US-Outbound-Travel-Country-Region-Historical-2008-2016.pdf; Russian Federation Federal State Statistics Service, "2017 Russian Statistical Yearbook—Inbound Tourist Visits of Foreign Citizens to Russia," available at www.gks.ru/free_doc/doc_2017/year/year17.pdf.
69. Kenneth Rapoza, "American Companies Thriving in Russia: Is There Something to Be Ashamed About?," *Forbes,* August 3, 2016; Alanna Petroff, U.S.-Russia Business Ties: 5 Things You Need to Know," CNN Money, July 7, 2017; Ilya Khrennikov, "Big Western Companies Are Pumping Cash into Russia," Bloomberg, November 22, 2016.
70. Bre Payton, "If Mere 'Contact' With Foreigners Is a Crime, Obama Should've Been Locked Up in 2008," *Federalist,* December 1, 2017.
71. Bill Kovach, "Salinger Talked to Foe on P.O.W.'s as McGovern Aide," *New York Times,* August 17, 1972.

72. Sean Davis, "Ted Kennedy Secretly Asked the Soviets to Intervene in 1984 Elections," *Federalist,* March 10, 2015.
73. Ned Parker, Jonathan Landay, aqnd Warren Strobel, "Exclusive: Trump Campaign Had at Least 18 Undisclosed Contacts with Russians: Sources," Reuters, May 18, 2017.
74. Ibid.
75. Olivia Beavers, "Kremlin Spokesman: Russian Ambassador Met with Advisers to Clinton Campaign Too," *The Hill,* March 12, 2017.
76. Matt Apuzzo, Adam Goldman, and Nicholas Fandos, "Code Name Crossfire Hurricane: The Secret Origins of The Trump Investigation," *New York Times,* May 16, 2018.
77. Adam Edelman, "DOJ Seeks Probe of FBI Conduct in 2016 Campaign after Trump 'Spy' Claim," NBC News, May 21, 2018.
78. Michael Crowley, "Trump's Foreign Policy Team Baffles GOP Experts," *Politico,* March 21, 2016; Philip Rucker and Robert Costa, "Donald Trump Names Foreign Policy Advisers," *Washington Post,* March 21, 2016.
79. Kaitlyn Schallhorn, "George Papadopoulos, ex-Trump Aide, Pleads Guilty in Connection to Russia Probe: Who Is He?," Fox News, January 2, 2018.
80. Sharon LaFraniere, Mark Mazzetti, and Matt Apuzzo, "How the Russia Inquiry Began: A Campaign Aide, Drinks and Talk of Political Dirt," *New York Times,* December 30, 2017.
81. Ibid.
82. Ibid.
83. Christopher Brennan, "Mysterious Professor in Papadopoulos Docs Says He Did Not Discuss Clinton Emails," *New York Daily News,* October 31, 2017.
84. Cristiano Lima, "6 Key Findings From Papadopoulos' Guilty Plea in Russia Probe," *Politico,* October 30, 2017; Rosalind S. Helderman and Tom Hamburger, "Top Campaign Officials Knew of Trump Adviser's Outreach to Russia," *Washington Post,* October 30, 2017.

CHAPTER 6: THE FABRICATED "DOSSIER" USED AGAINST TRUMP

1. Paul Sperry, "Exclusive: CIA Ex-Director Brennan's Perjury Peril," RealClear Investigations, February 11, 2018.
2. Ibid.
3. Greg Miller, "CIA Director Alerted FBI to Pattern of Contacts Between Russian Officials and Trump Campaign Associates," *Washington Post,* May 23, 2017; CNN Newsroom, "Transcript: Former CIA Director John Brennan's Testimony Before Congress," CNN, May 23, 2017, available at http://www.cnn.com/TRANSCRIPTS/1705/23/cnr.04.html.
4. Eric Lichtblau, "CIA Had Evidence of Russia Effort to Help Trump Earlier Than Believed," *New York Times,* April 6, 2017.
5. Amy Entous, Devlin Barrett, and Rosalind S. Helderman, "Clinton Campaign, DNC Paid for Research That Led to Russia Dossier," *Washington Post,* October 24, 2017.

6. Kevin Dilanian, "Mueller's Team Traveled to Interview Ex-Spy Involved in Dossier," NBC News, October 5, 2017.
7. Stephen F. Cohen, "Russiagate or Intelgate?," *Nation,* February 7, 2018, available at https://www.thenation.com/article/russiagate-or-intelgate/.
8. Interview of James Kallstrom with Maria Bartiromo, "Sunday Morning Futures," March 18, 2018; Tim Hains, "Kallstrom: Brennan and Obama Are the People Who Committed Felonies, Not General Flynn," RealClearPolitics, March 18, 2018.
9. Paul Sperry, "Exclusive: CIA Ex-Director Brennan's Perjury Peril," RealClear Investigations, February 11, 2018.
10. Ibid.
11. CNN Staff, "Testimony of CIA Director John Brennan before the House Intelligence Committee," CNN, Mary 23, 2017, available at http://www.cnn.com/TRANSCRIPTS/1705/23/cnr.04.html.
12. Lee Smith, "How CIA Director John Brennan Targeted James Comey," *Tablet,* February 9, 2018.
13. Sophie Tatum, "Former CIA Chief to Trump on McCabe Firing: 'America Will Triumph Over You,'" CNN, March 17, 2018.
14. Kimberly Leonard, "Samantha Power Delivers Ominous Message to Trump: Not A Good Idea to Piss Off John Brennan," *Washington Examiner*, March 17, 2018.
15. Byron York, "Did Dossier Trigger the Trump-Russia Probe?," *Washington Examiner,* November 12, 2017.
16. Letter from Senator Harry Reid to FBI Director James Comey, August 27, 2016, available at https://archive.org/details/ReidLetterToComey08272016.
17. House Permanent Select Committee on Intelligence, "Summary Table of Findings," p. 4, March 2018, available at https://intelligence.house.gov/uploadedfiles/russia_report_findings_and_recommendations.pdf.
18. Ian Schwartz, "GOP Rep. Jordan: House Intel Committee 'Confirmed' James Clapper Leaked Info on Russian Probe," RealClearPolitics, March 14, 2018.
19. Ibid.
20. House Permanent Select Committee on Intelligence, "Summary Table of Findings," p. 4, March 2018, available at https://intelligence.house.gov/uploadedfiles/russia_report_findings_and_recommendations.pdf.
21. Scott Shane, Nicholas Confessore, and Matthew Rosenberg, "How a Sensational, Unverified Dossier Became a Crisis for Donald Trump," *New York Times*, January 11, 2017; Amy Entous, Devlin Barrett, and Rosalind S. Helderman, "Clinton Campaign, DNC Paid for Research That Led to Russia Dossier," *Washington Post,* October 24, 2017.
22. Kenneth P. Vogel, "Clinton Campaign and Democratic Party Helped Pay for Russia Trump Dossier," *New York Times,* October 24, 2017; Mark Hosenball, "Ex-British Spy Paid $168,000 for Trump Dossier, U.S. Firm Discloses," Reuters, November 1, 2017.
23. Ken Bensinger, Miriam Elder, and Mark Schoofs, "These Reports Allege Trump Has Deep Ties to Russia," *Buzzfeed,* January 10, 2017, available at https://www

.buzzfeed.com/kenbensinger/these-reports-allege-trump-has-deep-ties-to-russia?utm_term=.nkzKndwXx#.bmqVQvK34.

24. Jeremy Herb, Manu Raju, and Marshall Cohen, "Fusion Co-Founder: Dossier Author Feared Trump Was Being Blackmailed," CNN, January 10, 2018.
25. Tom Hamburger and Rosalind S. Helderman, "Hero or Hired Gun? How A British Former Spy Became A Flash Point In The Russia Investigation," *The Washington Post*, February 6, 2018.
26. Ibid.
27. Lena Felton, "The Full Text of the Nunes Memo," *Atlantic,* February 2, 2018.
28. Ken Bensinger, Mirum Elder, and Mark Schoofs, "These Reports Allege Trump Has Deep Ties to Russia."
29. Paul Roderick Gregory, "The Trump Dossier Is Fake—And Here Are the Reasons Why," *Forbes,* January 13, 2017.
30. Ibid.
31. Owen Matthews, "Thirteen Things That Don't Add Up In The Russia-Trump Intelligence Dossier," *Newsweek*, January 11, 2017.
32. Ibid.
33. Rosie Gray, "Michael Cohen: 'It Is Fake News Meant to Malign Mr. Trump,'" *Atlantic,* January 10, 2017.
34. Ian Schwartz, "Woodward: Trump Right to Be Upset About 'Garbage Document,' Deserves Apology," RealClearPolitics, January 15, 2017.
35. Fox News Insider, "'Garbage Document': Woodward Says U.S. Intel Should Apologize Over Trump Dossier," Fox News, January 16, 2017.
36. Bill Neely, "Kremlin Spokesman: U.S. Intelligence Report on Russian Hacking 'Ridiculous,'" NBC News, January 12, 2017.
37. Fox News (Associate Press contributed), "Putin: Those Who Leaked Trump Dossier 'Worse Than Prostitutes,'" Fox News, January 17, 2017.
38. Ibid.
39. Anna Giaritelli, "Carter Page Says He's Never Spoken with Trump in His Life," *Washington Examiner*, February 6, 2018; Testimony of Carter Page Before the House Intelligence Committee, November 2, 2017, pp. 11 and 15, available at https://intelligence.house.gov/uploadedfiles/carter_page_hpsci_hearing_transcript_nov_2_2017.pdf.
40. Ibid., p. 11.
41. Ibid., p. 157.
42. Scott Shane, Mark Mazzetti, and Adam Goldman, "Trump Adviser's Visit to Moscow Got the FBI's Attention," *New York Times*, April 19, 2017.
43. Testimony of Carter Page Before the House Intelligence Committee, November 2, 2017, p. 19, available at https://intelligence.house.gov/uploadedfiles/carter_page_hpsci_hearing_transcript_nov_2_2017.pdf/.
44. Ibid., p. 40.
45. Manu Raju, Jeremy Herb, and Katelyn Polantz, "Carter Page Reveals New Contacts with Trump Campaign, Russians," CNN, November 8, 2017.
46. Ibid.
47. Ibid.

48. Testimony of Carter Page Before the House Intelligence Committee, November 2, 2017, p. 38, available at https://intelligence.house.gov/uploadedfiles/carter_page_hpsci_hearing_transcript_nov_2_2017.pdf.
49. Ibid., p. 36.
50. "Dossier" Purportedly Prepared by Christopher Steele, Orbis Business Intelligence, "Company Intelligence Report 2016," p. 9, available at https://www.documentcloud.org/documents/3259984-Trump-Intelligence-Allegations.html, published by *Buzzfeed*, January 10, 2017.
51. Ibid., p. 25.
52. Ibid., pp. 20–22.
53. "Dossier" purportedly prepared by Christopher Steele, Orbis Business Intelligence, "Company Intelligence Report 2016," p. 30, available at https://www.documentcloud.org/documents/3259984-Trump-Intelligence-Allegations.html.
54. Manu Raju, Jeremy Herb, and Katelyn Polantz, "Carter Page Reveals New Contacts with Trump Campaign, Russians," CNN, November 8, 2017.
55. Elena Mazneva and Ilya Arkhipov, "Russia Sells $11 Billion Stake in Rosneft to Glencore and Qatar," Bloomberg, December 7, 2016.
56. *Washington Post* Staff, "Full Transcript: FBI Director James Comey Testifies on Russian Interference in 2016 Election," *Washington Post*, March 20, 2017, available at https://www.washingtonpost.com/news/post-politics/wp/2017/03/20/full-transcript-fbi-director-james-comey-testifies-on-russian-interference-in-2016-election/?utm_term=.7c66b8617838.
57. Elena Mazneva and Ilya Arkhipov, "Russia Sells $11 Billion Stake in Rosneft to Glencore And Qatar."
58. Testimony of Carter Page Before the House Intelligence Committee, November 2, 2017, pp. 7–8, available at https://intelligence.house.gov/uploadedfiles/carter_page_hpsci_hearing_transcript_nov_2_2017.pdf.
59. Elena Mazneva and Ilya Arkhipov, "Russia Sells $11 Billion Stake in Rosneft to Glencore and Qatar."
60. Margarita Papchenkova, "Rosneft Looking for Investors in Itself," *Corporation Energy*, July 11, 2016.
61. Andrew C. McCarthy, "The Schiff Memo Harms Democrats More Than It Helps Them," *National Review*, February 25, 2018; United States Code, 50 U.S.C. 1804, "Applications for Court Orders."

CHAPTER 7: GOVERNMENT ABUSE OF SURVEILLANCE

1. *Illinois v. Gates*, 462 U.S. 213 (1983).
2. Kenneth P. Vogel and Maggie Haberman, "Conservative Website First Funded Anti-Trump Research by Firm That Later Produced Dossier," *New York Times*, October 27, 2017.
3. Scott Shane, Nicholas Confessore, and Matthew Rosenberg, "How a Sensational, Unverified Dossier Became a Crisis for Donald Trump," *New York Times*, January 11, 2017.
4. Senate Judiciary Committee, U.S. Senate, Interview of Glenn Simpson, Founder Fusion GPS, August 22, 2017, p. 148, available at https://www.feinstein.senate

.gov/public/_cache/files/3/9/3974a291-ddbe-4525-9ed1-22bab43c05ae/934A3562824CACA7BB4D915E97709D2F.simpson-transcript-redacted.pdf.

5. Ibid., p. 93.
6. Ibid., p. 233.
7. Ibid., p. 235.
8. Ibid., p. 246.
9. Ibid., p. 279.
10. Ibid., p. 203.
11. Ibid., p. 275.
12. Ibid., p. 292.
13. Ibid., p. 170.
14. Ibid., p. 242.
15. Kimberley A. Strassel, "Who Paid for the 'Trump Dossier'?," *Wall Street Journal*, July 27, 2017.
16. Kimberley A. Strassel, "The Fusion Collusion," *Wall Street Journal*, October 19, 2017.
17. Ibid.
18. Senate Judiciary Committee, U.S. Senate, Interview of Glenn Simpson, Founder Fusion GPS, August 22, 2017, pp. 160–61, available at https://www.feinstein.senate.gov/public/_cache/files/3/9/3974a291-ddbe-4525-9ed1-22bab43c05ae/934A3562824CACA7BB4D915E97709D2F.simpson-transcript-redacted.pdf.
19. Ibid., p. 210.
20. Todd Shepherd, "Fusion GPS Paid Journalists, Court Papers Confirm," *Washington Examiner*, November 21, 2017.
21. Ibid.
22. Ibid.
23. House Permanent Select Committee on Intelligence, "Memorandum: Foreign Intelligence Surveillance Act Abuses at the Department of Justice and the Federal Bureau of Investigation," January 18, 2018 (declassified by order of the president, February 2, 2018), available at https://intelligence.house.gov/uploadedfiles/memo_and_white_house_letter.pdf.
24. James Rosen and Jake Gibson, "Wife of Demoted DOJ Official Worked for Firm Behind Anti-Trump Dossier," Fox News, December 11, 2017.
25. James Rosen and Jake Gibson, "Top DOJ Official Demoted Amid Probe of Contacts with Trump Dossier Firm," Fox News, December 7, 2017.
26. Ibid.
27. Luke Rosiak, "DOJ Official Bruce Ohr Hid Wife's Fusion GPS Payments from Ethics Officials," *Daily Caller*, February 14, 2018.
28. United States Code, 18 U.S.C. 1001, "Statements or Entries Generally."
29. Rowan Scarborough, "Hillary Defends Trump Dossier, Makes Dubious Claims About Its Release," *Washington Times*, November 2, 2017.
30. Ned Ryun, "There Is Nothing Normal About the Fusion GPS Dossier," *The Hill*, November 5, 2017.
31. Ariella Phillips, "Former Trump Adviser Carter Page Under FISA Warrant Since 2014: Report," *Washington Examiner*, August 3, 2017.
32. Rebecca Ballhaus and Byron Tau, "Former Trump Aide Carter Page Was on U.S.

Counterintelligence Radar Before Russia Dossier," *Wall Street Journal*, February 1, 2018.

33. Andrew C. McCarthy, "A Foreign Power's Recruitment Effort Is Not a Basis for a FISA Court Warrant," *National Review*, February 26, 2018.
34. House Permanent Select Committee on Intelligence, "Memorandum: Foreign Intelligence Surveillance Act Abuses at the Department of Justice and the Federal Bureau of Investigation," January 18, 2018 (Declassified by order of the President, February 2, 2018), available at https://intelligence.house.gov/uploadedfiles/memo_and_white_house_letter.pdf.
35. Dartunorrow Clark, "Schumer, Pelosi Call for Nunes to Be Removed as House Intelligence Chair," NBC News, February 1, 2018.
36. Ibid.
37. CBS News, "Republicans Call for Release of Memo on Alleged Surveillance Abuses," January 19, 2018.
38. Shimon Prokupecz, Laura Jarrett, Jim Sciutto, Abby Phillip, and Kevin Liptak, "FBI Chief Has 'Grave Concerns,' Clashes with Trump Over GOP Memo," CNN, February 1, 2018.
39. Nicholas Fandos, "Committee Votes to Release Democratic Rebuttal to G.O.P. Russia Memo," *New York Times*, February 5, 2018.
40. House Permanent Select Committee on Intelligence, Minority, "Memo: Correcting the Record—The Russia Investigations," *Unclassified*, p. 1, January 29, 2018 (released to the public February 24, 2018).
41. Andrew C. McCarthy, "The Schiff Memo Harms Democrats More Than It Helps Them," *National Review*, February 25, 2018; United States Code, 50 U.S.C. 1804, "Applications for Court Orders."
42. House Permanent Select Committee on Intelligence, Minority, "Memo: Correcting the Record—The Russia Investigations," *Unclassified*, p. 5, January 29, 2018 (released to the public February 24, 2018).
43. Ibid., p. 4.
44. Spencer Ackerman, "Sources: Devin Nunes Memo Is 100% Wrong About Andrew McCabe and Steele Dossier," *The Daily Beast*, February 2, 2018; Diana Stancy Correll, "Doubt Cast Upon Nunes Memo's Bombshell Claim About Andrew McCabe's Testimony on Trump Dossier," *Washington Examiner*, February 2, 2018.
45. Fred Fleitz, "Democrats' New Russia Memo Is an Anti-Trump Political Attack," Fox News, February 25, 2018.
46. James B. Comey, "Statement for the Record," Senate Select Committee on Intelligence, June 8, 2017, available at https://www.scribd.com/document/350651927/James-Comey-s-Opening-Statement?irgwc=1&content=10079&campaign=Skimbit%2C%20Ltd.&ad_group=58287X1517246Xecc7c2adb69f57650dc89af0db896eb0&keyword=ft750noi&source=impactradius&medium=affiliate#from_embed.
47. James Rosen and Jake Gibson, "McCabe Draws Blank on Democrats' Funding of Trump Dossier, New Subpoenas Planned," Fox News, December 20, 2017; Rowan Scarborough, "Embattled FBI Admits It Can't Verify Dossier Claims of Russia, Trump Campaign Collusion," *Washington Times*, December 25, 2017.

48. Letter and Memorandum from Senator Charles E. Grassley and Senator Lindsey O. Graham, "Referral of Christopher Steel for Potential Violation of 18 U.S.C. 1001," January 4, 2018 (made public February 10, 2018), available at https://www.judiciary.senate.gov/imo/media/doc/2018-02-06%20CEG%20LG%20to%20DOJ%20FBI%20(Unclassified%20Steele%20Referral).pdf.
49. Ibid., p. 3.
50. House Permanent Select Committee on Intelligence, Minority, "Memo: Correcting the Record—The Russia Investigations," *Unclassified*, p. 5, January 29, 2018 (released to the public February 24, 2018).
51. Larry Abramson, "FISA Court: We Approve 99% of Wiretap Applications," National Public Radio, October 15, 2013.
52. Ibid., p. 4.
53. Ibid., pp. 3 and 4.
54. Letter and Memorandum from Senator Charles E. Grassley, Senator Lindsey O. Graham, Senator John Cornyn, and Senator Thom Tillis, "Referral of Christopher Steele for Potential Violation of 18 U.S.C. 1001," March 15, 2018.
55. Ibid.
56. Ibid.
57. In the High Court of Justice Queen's Bench Division, *Gubarev v. Orbis,* "Defendants' Response to Claimants' Request for Further Information Pursuant to CPR Part 18," p. 7, available at https://assets.documentcloud.org/documents/3892131/Trump-Dossier-Suit.pdf.
58. Ibid., p. 7.
59. United States Code, 18 U.S.C. 1001, "Statements or Entries Generally."
60. Letter and Memorandum from Senator Charles E. Grassley and Senator Lindsey O. Graham, "Referral of Christopher Steel for Potential Violation of 18 U.S.C. 1001," January 4, 2018 (made public February 10, 2018), p. 7, available at https://www.judiciary.senate.gov/imo/media/doc/2018-02-06%20CEG%20LG%20to%20DOJ%20FBI%20(Unclassified%20Steele%20Referral).pdf.
61. Foreign Intelligence Surveillance Act, 50 U.S.C. Chapter 36, section 1801 et seq.
62. James Rosen and Jake Gibson, "McCabe Draws Blank on Democrats' Funding of Trump Dossier, New Subpoenas Planned," Fox News, December 20, 2017; Rowan Scarborough, "Embattled FBI Admits It Can't Verify Dossier Claims of Russia, Trump Campaign Collusion," *Washington Times*, December 25, 2017.
63. *Weeks v. United States*, 232 U.S. 383 (1914); *Mapp v. Ohio*, 367 U.S. 643 (1961); *Silverthorne Lumber Co. v. United States*, 251 U.S. 385 (1920).
64. *Nardone v. United States,* 308 U.S. 338 (1939).
65. United States Code, 18 U.S.C. 242, "Deprivation of Rights Under Color of Law."
66. United States Department of Justice, "Deprivation of Rights Under Color of Law, available at https://www.justice.gov/crt/deprivation-rights-under-color-law.
67. Ibid.
68. Ibid.
69. United States Code, 18 U.S.C. 1621, "Perjury Generally"; United States Code, 18 U.S.C. 1623, "False Declarations Before Grand Jury or Court."

70. Ibid.
71. United States Code, 18 U.S.C. 1001, "Statements or Entries Generally."
72. Ibid.
73. United States Code, 18 U.S.C. 1503, "Influencing or Injuring Officer or Juror Generally."
74. United States Code, 18 U.S.C. 1515(b), "Definitions for Certain Provisions; General Provision."
75. United States Code, 18 U.S.C. 1031, "Major Fraud Against the United States."
76. Ibid.
77. United States Code, 18 U.S.C. 371, "Conspiracy to Commit Offense or To Defraud United States."
78. Ibid.
79. United States Code, 18 U.S.C. 1001, "Statements or Entries Generally."
80. United States Code, 18 U.S.C. 951, "Agents of Foreign Governments"; United States Code, 22 U.S.C. 611 et seq., "Foreign Relations and Intercourse."
81. Letter from Sen. Charles E. Grassley, chairman, Committee on the Judiciary Committee, to Dana Boente, acting deputy attorney general, March 31, 2017.
82. United States Code, 52 U.S.C. 30121, "Contributions and Donations by Foreign Nationals."
83. United States Code, 52 U.S.C. 30101 et seq., "Voting and Elections."
84. Complaint Filed Before The Federal Election Commission, October 25, 2017 (filed by Campaign Legal Center); http://www.campaignlegalcenter.org/document/fec-complaint-hillary-america-dnc-failure-disclose.
85. Ciara Torres-Spelliscy, "The Justice Department Is Now on the Campaign Beat," Brennan Center for Justice, New York University School of Law, October 12, 2015; Crag Donsanto and Nancy Simmons, "Federal Prosecution of Election Offenses" (Seventh Edition), Department of Justice, May 2007, available at https://www.justice.gov/sites/default/files/criminal/legacy/2013/09/30/electbook-rvs0807.pdf.

CHAPTER 8: MEETING WITH RUSSIANS IS NOT A CRIME

1. Joe Becker, Matt Apuzzo, and Adam Goldman, "Trump Team Met with Lawyer Linked to Kremlin During Campaign," *New York Times*, July 8, 2017.
2. Tom Porter, "Donald Trump Jr. 'Treason' Emails Prove Russian Collusion: Tim Kaine," *Newsweek,* July 11, 2017.
3. Rachel Stockman, "'This Is Treason': Some Legal Experts Say Trump Jr.'s Clinton Dirt/Russia Meeting Was Illegal," LawNewz.com, July 9, 2017.
4. Ibid.
5. Constitution of the United States of America, Article III, Section 3; United States Code, 18 U.S.C. 2381, "Treason."
6. Gabriel Schoenfield, "Donald Trump Jr. and the Whiff of Treason: Morally, He's in Deep," *USA Today,* July 19, 2017.
7. Carlton F. W. Larson, "Five Myths About Treason," *Washington Post,* February 17, 2017.
8. Ibid.

9. United States Code, 18 U.S.C. 371, "Conspiracy to Commit Offense or to Defraud United States."
10. *Hass v. Henkel,* 216 U.S. 462 (1910); *Hammerschmidt v. United States,* 265 U.S. 182. See "Conspiracy to Defraud the United States," U.S. Department of Justice, Offices of the United States Attorneys, available at https://www.justice.gov/usam/criminal-resource-manual-923-18-usc-371-conspiracy-defraud-us.
11. *Hammerschmidt,* p. 188.
12. Jo Becker, Adam Goldman, and Matt Apuzzo, "Russian Dirt on Clinton? 'I Love It,' Donald Trump Jr. Said," *New York Times,* July 11, 2017.
13. Catherine Herridge, Pamela K. Browne, and Cyd Upson, "Russian Lawyer at Center of Trump Tower Meeting Dismisses Dossier Shared With FBI," Fox News, January 19, 2018.
14. Ibid.
15. Karoun Demirjian, "Pelosi Suggests Trump Surrogates Violated Law as Members Try to Force Votes on Matters Related to Russia Probe," *Washington Post,* July 14, 2017.
16. Federal Election Commission, "Volunteer Activity," available at https://transition.fec.gov/pages/brochures/volact.shtml.
17. Ibid.
18. United States Code, 52 U.S.C. 30121, "Contributions and Donations by Foreign Nationals."
19. Code of Federal Regulations, 11 C.F.R. 110.20, "Prohibition on Contributions, Donations, Expenditures, Independent Expenditures, and Disbursements by Foreign Nationals."
20. Federal Election Commission, "Volunteer Activity," available at https://transition.fec.gov/pages/brochures/volact.shtml.
21. Code of Federal Regulations, 11 C.F.R. 100.74, "Uncompensated Services by Volunteers."
22. Jonathan Turley, "Don Jr.'s Russia Meeting Wasn't Collusion—Just Amateur Hour," *The Hill,* July 11, 2017.
23. Robert Barnes, "If Trump Jr. Is Guilty, So Is Every Democrat Who Takes Information From 'Dreamers,'" *Law and Crime,* July 17, 2017, available at https://lawandcrime.com/important/if-donald-trump-jr-is-guilty-so-is-every-democrat-who-takes-information-from-a-dreamer/.
24. United States Code, 52 U.S.C. 30109(d)(1)(A), "Federal Election Campaign Act: Enforcement," available at http://uscode.house.gov/view.xhtml?req=(title:52%20section:30109%20edition:prelim); Andy Grewal, "If Trump Jr. Didn't Know Campaign Finance Law, He Won't Be Prosecuted," Notice & Comment, *Yale Journal on Regulation,* July 16, 2017, available at http://yalejreg.com/nc/if trump-jr-didnt-know-campaign-finance-law-he-didnt-break-it/.
25. Ellen Nakashima, Adam Entous, and Greg Miller, "Russia Ambassador Told Moscow That Kushner Wanted Secret Communications Channel with Kremlin," *Washington Post,* May 26, 2017.
26. Ibid.
27. Robert F. Kennedy, *Thirteen Days: A Memoir of the Cuban Missile Crisis* (New York: Norton, 1969).

28. Abby Phillip, "Homeland Security Chief Defends Kushner's Alleged Proposal for 'Back Channel' to the Russians as 'A Good Thing,'" *Washington Post*, May 28, 2017.
29. Riley Beggin, "Back-Channel Communications Are Nothing New for White House," ABC News, May 31, 2017.
30. Statement of Jared Kushner to Congressional Committees, July 24, 2017, p. 10, available at http://www.foxnews.com/politics/interactive/2017/07/24/jared-kushner-statement-to-congressional-committees.html.
31. Ibid., p. 7.
32. Ibid., p. 8.
33. Interview with Professor W. David Clinton, chairman of the Political Science Department at Baylor University, July 24, 2017.
34. Ibid.
35. United States Code, 18 U.S.C. 953, "Private Correspondence with Foreign Governments" (The Logan Act).
36. Statement of Jared Kushner to Congressional Committees, July 24, 2017, p. 7. Available at http://www.foxnews.com/politics/interactive/2017/07/24/jared-kushner-statement-to-congressional-committees.html.
37. United States Code, 18 U.S. C. 1001, "Statements of Entries Generally"; see also Questionnaire for National Security Positions, Standard Form 86, U.S. Office of Personnel Management (5 C.F.R. 731–736), available at https://www.opm.gov/forms/pdf_fill/sf86-non508.pdf.
38. Statement of Jared Kushner to Congressional Committees, July 24, 2017, p. 10. Available at http://www.foxnews.com/politics/interactive/2017/07/24/jared-kushner-statement-to-congressional-committees.html.
39. Questionnaire for National Security Positions, Standard Form 86C, U.S. Office of Personnel Management (5 C.F.R. 731–736), available at https://www.opm.gov/Forms/pdf_fill/SF86C.pdf.
40. *Cox v. Louisiana*, 379 U.S. 559 (1965), available at https://www.law.cornell.edu/supremecourt/text/379/559.
41. Representative Nancy Pelosi, "Pelosi Statement on Attorney General Sessions' Apparent Perjury on Russian Meetings," Press Release, March 1, 2017, available at https://pelosi.house.gov/news/press-releases/pelosi-statement-on-attorney-general-sessions-apparent-perjury-on-russian.
42. Alana Abramson, "Here's Exactly What Jeff Sessions Said About Russia at His Confirmation Hearing," *Time,* March 2, 2017.
43. Lauren Carroll, "In Context: What Jeff Sessions Told Al Franken About Meeting Russian Officials," PolitiFact, March 2, 2017, available at http://www.politifact.com/truth-o-meter/article/2017/mar/02/context-what-jeff-sessions-told-al-franken-about-m/.
44. Fox News, "Sessions, Russian Ambassador Reportedly Spoke Twice During Presidential Campaign," Fox News, March 2, 2017.
45. Aaron Blake, "Transcript of Jeff Sessions's Recusal News Conference, Annotated," *Washington Post*, March 2, 2017.
46. Attorney General Jeff Sessions, "Prepared Remarks to the United States Senate Select Committee on Intelligence," Department of Justice, June 13, 2017.

47. Aaron Blake, "Transcript of Jeff Sessions's Recusal News Conference, Annotated."
48. Ibid.
49. Deirdre Shesgreen, "Claire McCaskill Accused of Lying in Attack on Jeff Sessions," *USA Today,* March 2, 2017.
50. Fox News, "Sessions Not Alone: Russian Ambassador Also Met with Numerous Democrats," Fox News, March 3, 2017.
51. Kaitlan Collins, "Russian Ambassador Sergey Kislyak Appeared as Obama White House Visitor at Least 22 Times," *Daily Caller,* March 2, 2017.
52. NBC News, Trump Tweets Old Photos of Schumer, Pelosi with Russian Leaders," NBC News, March 3, 2017.
53. United States Code, 18 U.S.C. 1621, "Perjury Generally."
54. *Politico* Staff, "Transcript: Jeff Sessions' Testimony on Trump and Russia," *Politico,* June 13, 2017.
55. Ibid.
56. Ibid.
57. Manu Raju and Evan Perez, "First on CNN: AG Sessions Did Not Disclose Russia Meetings in Security Clearance Form, DOJ says," CNN, May 25, 2017.
58. Brian Flood, "CNN Quietly Backtracks Another Report Tying Trump Campaign to Russia," Fox News, December 11, 2017.

CHAPTER 9: FLYNN'S FIRING, SESSIONS'S RECUSAL, AND THE CANNING OF COMEY

1. Greg Miller and Adam Goldman, "Head of Pentagon Intelligence Agency Forced Out, Officials Say," *Washington Post*, April 30, 2014.
2. Ibid.
3. Chuck Ross, "Here's What Prompted Michael Flynn to Register as an Agent of Turkey," *Daily Caller*, March 31, 2017.
4. David S. Cloud and Christine Mai-Duc, "Retired Army Gen. Michael Flynn Delivers Fiery Speech to Emptying Convention Hall," *Los Angeles Times,* July 20, 2016; Stephen Dinan, "Michael Flynn Calls for Hillary Clinton to Quit Race, Go to Prison," *Washington Times*, July 18, 2016.
5. Edward-Isaac Dovere and Matthew Nussbaum, "Obama Warned Trump About Flynn, Officials Say," *Politico*, May 8, 2017.
6. Michael D. Shear, "Obama Warned Trump About Hiring Flynn, Officials Say," *New York Times*, May 8, 2017.
7. Rich Lowry, "No, Michael Flynn Didn't Violate the Logan Act," *National Review*, December 1, 2017.
8. David E. Sanger, "Obama Strikes Back at Russia for Election Hacking," *New York Times*, December 29, 2016.
9. Statement of The Offense, *U.S. v. Michael Flynn*, United States District Court for The District of Columbia, November 30, 2017.
10. Letter of Senator Charles Grassley, chairman of the Committee on the Judiciary, May 11, 2018, https://www.judiciary.senate.gov/imo/media/doc/2018-05-11%20CEG%20to%20DOJ%20FBI%20(Flynn%20Transcript).pdf.

11. Statement of The Offense, *U.S. v. Michael Flynn*, United States District Court for The District of Columbia, November 30, 2017.
12. United States Code, 18 U.S.C. 1001, "Statements or Entries Generally."
13. Nicole Darrah, "Michael Flynn Selling Home to Pay for Legal Fees After Pleading Guilty in Trump Probe," Fox News, March 6, 2018.
14. The Editorial Board, "The Flynn Information," *Wall Street Journal*, December 1, 2017.
15. Jim Sciutto and Marshall Cohen, "Flynn Worries About Son in Special Counsel Probe," CNN, November 9, 2017.
16. United States Code, 18 U.S.C. 1001, "Statements or Entries Generally."
17. Evan Perez, "Flynn Changed Story to FBI, No Charges Expected," CNN, February 17, 2017.
18. The Editorial Board, "The Flynn Information," *Wall Street Journal*, December 1, 2017.
19. Andrew C. McCarthy, "On Strzok, Let's Wait for the Evidence," *National Review*, December 6, 2017; Andrew C. McCarthy, "The Curious Michael Flynn Guilty Plea," *National Review*, February 13, 2018.
20. Byron York, "Comey Told Congress FBI Agents Didn't Think Michael Flynn Lied," *Washington Examiner*, February 12, 2018.
21. *Brady v. Maryland*, 373 U.S. 83 (1963); See also *United States v. Agurs*, 427 U.S. 97, 107 (1976).
22. Mollie Hemingway, "Revealed: Peter Strzok Had Personal Relationship with Recused Judge in Michael Flynn Case," *Federalist,* March 16, 2018; Sara A. Carter, "New Text Messages Reveal FBI Agent Was Friends with Judge in Flynn Case," March 21, 2018, available at https://saraacarter.com/explosive-text-messages-reveal-judge-in-flynn-case-was-friends-with-strzok/.
23. Ibid.
24. Mollie Hemingway, "Revealed: Peter Strzok Had Personal Relationship with Recused Judge in Michael Flynn Case."
25. Nedra Pickler, "Justice Department Lawyers in Contempt for Withholding Stevens Documents," Associated Press, February 14, 2009.
26. Erika Bolstad and Richard Maur, "U.S. Attorney General Ends Stevens Prosecution," *Anchorage Daily News*, April 1, 2019.
27. Order of the U.S. District Court for The District of Columbia (Criminal No. 17-232-10 EGS), Judge Emmet G. Sullivan, pp. 2 and 3, February 16, 2018.
28. Ibid.
29. United States Code, 18 U.S.C. 953, "Private Correspondence with Foreign Governments" (The Logan Act).
30. *Waldron v. British Petroleum Co.*, 231 Fed. Supp. 72 (1964).
31. Statement of the Offense, *U.S. v. Michael Flynn,* United States District Court for The District of Columbia, November 30, 2017.
32. Donald J. Trump, President, Twitter, December 22, 2016, available at https://twitter.com/realDonaldTrump/status/811928543366148096.
33. CBS/Associated Press, "U.N. Security Council Votes 14-0 to Condemn Israeli Settlement Construction," CBS News, December 23, 2016.

34. The Constitution of the United States, Article II, Section 3.
35. Byron York, "Comey Told Congress FBI Agents Didn't Think Michael Flynn Lied."
36. Ibid.
37. *Washington Post* Staff, "Full Transcript: Sally Yates and James Clapper Testify on Russian Election Interference," *Washington Post*, May 8, 2017.
38. Ibid.
39. Ibid.
40. Adam Shaw, "Shock Claim About FBI's Michael Flynn Interview Raises Questions," Fox News, February 14, 2018.
41. Transcript of "Meet the Press," NBC News, April 29, 2018, https://www.nbcnews.com/meet-the-press/meet-press-april-29-2018-n869906; transcript of "Special Report with Bret Baier," "James Comey on Clinton Probe, Russia Investigation," Fox News, April 26, 2018, http://www.foxnews.com/transcript/2018/04/26/james-comey-on-clinton-probe-russia-investigation.html.
42. House Permanent Select Committee on Intelligence, "Report on Russian Active Measures," p. 54, March 22, 2018, https://intelligence.house.gov/uploadedfiles/final_russia_investigation_report.pdf.
43. Letter by Senator Charles Grassley, chairman of Senate Committee on the Judiciary, May 11, 2018; https://www.judiciary.senate.gov/imo/media/doc/2018-05-11%20CEG%20to%20DOJ%20FBI%20(Flynn%20Transcript).pdf.
44. David Ignatius, "Why Did Obama Dawdle on Russia's Hacking?," *Washington Post*, January 12, 2017.
45. United States Code, 18 U.S.C. 798, "Disclosure of Classified Information."
46. *Washington Post* Staff, "Full Transcript: Sally Yates and James Clapper Testify on Russian Election Interference."
47. Greg Miller, Adam Entous, and Ellen Nakashima, "National Security Adviser Flynn Discussed Sanctions with Russian Ambassador, Despite Denials, Officials Say," *Washington Post*, February 9, 2017.
48. *Politico* Staff, "Transcript: Jeff Sessions' Testimony on Trump and Russia," *Politico*, June 13, 2017.
49. Eliza Relman, "Jeff Sessions Explains Why He Recused Himself from Trump Campaign-Related Investigations," *Business Insider*, June 13, 2017.
50. Fox News Staff, "Transcript: What Is Trump's Endgame with Sessions?," Fox News, July 25, 2017.
51. *Politico* Staff, "Transcript: Jeff Sessions' Testimony on Trump and Russia."
52. *Washington Post* Staff, "Full Transcript: FBI Director James Comey Testifies on Russian Interference in 2016 Election," *Washington Post*, March 2017.
53. Code of Federal Regulations, 28 C.F.R. 45.2, "Disqualification Arising from Personal or Political Relationship."
54. *Washington Post* Staff, "Full Transcript: FBI Director James Comey Testifies on Russian Interference In 2016 Election."
55. Andrew C. McCarthy, "Attorney General Sessions's Recusal Was Unnecessary," *National Review*, June 13, 2017; Andrew C. McCarthy, "A Special Prosecutor . . . for What?," *National Review*, March 3, 2017.

56. Matt Zapotosky, "Attorney General Declines to Provide Any Details on Clinton Email Investigation," *Washington Post*, July 12, 2017.
57. Michael S. Schmidt and Maggie Haberman, "Trump Humiliated Jeff Sessions After Mueller Appointment," *New York Times*, September 14, 2017.
58. Ibid.; Kelly Cohen, "Jeff Sessions Sent Trump a Resignation Letter in May After Humiliating Oval Office Meeting: Report," *Washington Examiner*, September 14, 2017.
59. Hearing Before the Committee on the Judiciary, "Oversight of the Federal Bureau of Investigation," Testimony of James B. Comey, September 28, 2016, available at https://judiciary.house.gov/wp-content/uploads/2016/09/114-91_22125.pdf; *Washington Post* Staff, "Read the Full Testimony of FBI Director James Comey In Which He Discusses Clinton Email Investigation, *Washington Post*, May 3, 2017.
60. Memorandum for the Attorney General, "Restoring Public Confidence in the FBI," Rod Rosenstein, deputy attorney general, May 9, 2017.
61. Ibid.
62. Ibid.

CHAPTER 10: OBSTRUCTION OF JUSTICE

1. "Rod Rosenstein Full Remarks to Congress on Comey Memo," *Axios*, May 19, 2017.
2. *Politico* Staff, "Full Text: James Comey Testimony Transcript on Trump and Russia," p. 32, *Politico*, June 8, 2017.
3. James B. Comey, Confidential Presidential Memorandum, FBI, January 28, 2017; Peter Kasperowicz, "James Comey Promised Trump: 'I Don't Leak,'" *Washington Examiner*, April 19, 2018.
4. *Politico* Staff, "Full Text: James Comey Testimony on Trump and Russia," *Politico*, June 8, 2017.
5. Ibid.
6. Ibid.
7. Michael S. Schmidt, "Comey Memo Says Trump Asked Him to End Flynn Investigation," May 16, 2017; David Lauter, "Trump Asked Comey to Shut Down Investigation of Flynn, *New York Times* Reports," *Los Angeles Times*, May 16, 2017.
8. Ibid., *New York Times*.
9. John Wagner, "'I Never Asked Comey to Stop Investigating Flynn': Trump Goes on Tweetstorm About the FBI," *Washington Post*, December 3, 2017.
10. Ibid.
11. United States Code, 18 U.S.C. 1505, "Obstruction of Proceedings Before Departments, Agencies, And Committees."
12. United States Code, 18 U.S.C. 1515(b), "Definitions for Certain Provisions; General Provision."
13. *Arthur Andersen v. United States*, 544 U.S. 696 (2005).
14. Interview with Fred Tecce, assistant U.S. attorney for the Eastern District of Pennsylvania and Special Assistant U.S. Attorney at the Department of Justice, April 1, 2018.

15. James B. Comey, "Statement for the Record," Senate Selection Committee on Intelligence, p. 5, June 8, 2017.
16. *Politico* Staff, "Full Text: James Comey Testimony on Trump and Russia," p. 20, *Politico,* June 8, 2017.
17. Ibid., p. 20.
18. Ibid., p. 9.
19. United States Code, 18 U.S.C. 4, "Misprision of Felony."
20. *Neal v. United States,* 102 F.2d 643 (1939) at 649; see also *Sullivan v. United States,* 411 F2d 556 (1969) and *United States v. Daddano,* 432 F2d 1119 (1970).
21. James B. Comey, "Statement for the Record," Senate Selection Committee on Intelligence, p. 5, June 8, 2017.
22. *Politico* Staff, "Full Text: James Comey Testimony on Trump and Russia," p. 66, *Politico,* June 8, 2017.
23. Ibid.
24. Constitution of the United States of America, Article II, Section 1.
25. Alan Dershowitz, "Trump Well Within Constitutional Authority on Comey, Flynn—Would This Even Be a Question If Hillary Were President?," Fox News, June 12, 2017.
26. Chris Wallace, interview with President Barack Obama, *Fox News Sunday,* Fox News, April 10, 2016.
27. Statement by FBI Director James B. Comey on the Investigation of Secretary Hillary Clinton's Use of a Personal Email System, FBI National Press Office, July 5, 2016.
28. Constitution of the United States of America, Article II, Section 3.
29. Andrew Prokop, "3 Potential Problems for an Obstruction of Justice Case Against Trump," Vox, January 25, 2018.
30. U.S. Citizenship and Immigration Services, "2014 Executive Actions on Immigration," available at https://www.uscis.gov/archive/2014-executive-actions-immigration.
31. Interview with Doug Burns, former assistant U.S. attorney for the Eastern District of New York, March 30, 2018.
32. Joseph Hinks, "Read Former FBI Director James Comey's Farewell Letter to Colleagues," *Time,* May 11, 2017.
33. Memorandum for the Attorney General, "Restoring Public Confidence in the FBI," Rod Rosenstein, deputy attorney general, May 9, 2017.
34. Jeff Sessions, Office of the Attorney General, Correspondence to President Donald Trump, May 9, 2017.
35. William Cummings, "Full Text of Trump's Letter Telling Comey He's Fired," *USA Today,* May 9, 2017.
36. "Rod Rosenstein Full Remarks to Congress on Comey Memo," *Axios.*
37. Ibid.
38. *Washington Post* Staff, "Full Transcript: Acting FBI Director McCabe and Others Testify Before Senate Intelligence Committee," *Washington Post,* May 11, 2017.
39. "Rod Rosenstein Full Remarks to Congress on Comey Memo," *Axios.*

40. *Washington Post* Staff, "Read Full Testimony of FBI Director James Comey in Which He Discusses Clinton Email Investigation," *Washington Post*, May 3, 2017.
41. James B. Comey, "Statement for the Record," Senate Selection Committee on Intelligence, p. 5, June 8, 2017.
42. Theresa Welsh and Matt Schofield, "Intel Officials Refuse to Answer Questions on Trump and Russia," McClatchy Company, June 7, 2017.
43. Ibid.
44. Tim Haines, "President Trump's Full Interview with Lester Holt: Firing of James Comey," p. 2, RealClearPolitics, May 11, 2017.
45. Ibid., pp. 3 and 4.
46. Ibid., p. 4.
47. Ibid., pp. 4 and 5.
48. Ibid., pp. 6 and 7.
49. Ibid., pp. 7 and 8.
50. Ibid., p. 8.
51. Brooke Singman, "DOJ Turns Over Redacted Comey Memos to Congressional Committees," Fox News, April 19, 2018; Samuel Chamberlain, "Trump Tweets That Comey Memos 'Clearly' Show No Obstruction, Collusion," Fox News, April 20, 2018.
52. *Politico* Staff, "Full Text: James Comey Testimony Transcript on Trump and Russia," pp. 8 and 9, *Politico*, June 8, 2017.
53. Matt Apuzzo, Maggie Haberman, and Matthew Rosenberg, "Trump Told Russians That Firing 'Nut Job' Comey Eased Pressure from Investigation," *New York Times*, May 19, 2017.
54. David S. Cloud, "Mueller Calls Back at Least One Participant in Key Meeting with Russians at Trump Tower," *Los Angeles Times*, January 6, 2018.
55. Ashley Parker, Carol D. Leonnig, Philip Rucker, and Tom Hamburger, "Trump Dictated Son's Misleading Statement on Meeting with Russian Lawyer," *Washington Post*, July 31, 2017; Liam Stack, "Donald Trump Jr.'s Two Different Explanations for Russian Meeting," *New York Times*, July 9, 2017.
56. Jo Becker, Matt Apuzzo, and Adam Goldman, "Trump's Son Met with Russian Lawyer After Being Promised Damaging Information on Clinton," *New York Times*, July 9, 2017.
57. Catherine Herridge, Pamela K. Browne, and Cyd Upson, "Russian Lawyer at Center of Trump Tower Meeting Dismisses Dossier Shared With FBI," Fox News, January 19, 2018.
58. Interview with Doug Burns, former assistant U.S. attorney for the Eastern District of New York, March 30, 2018.
59. *Politico* Staff, "Full Text: James Comey Testimony Transcript on Trump and Russia," pp. 32 and 33, *Politico*, June 8, 2017.
60. Michael S. Schmidt, "Comey Memo Says Trump Asked Him to End Flynn Investigation," *New York Times*, May 16, 2017.
61. James B. Comey, "Statement for the Record," Senate Select Committee on Intelligence, p. 5, June 8, 2017; *Politico* Staff, "Full Text: James Comey Testimony Transcript on Trump and Russia," *Politico*, June 8, 2017.

62. Federal Bureau of Investigation, "Manual of Investigative Operations and Guidelines (MIOG)," available at https://vault.fbi.gov/miog/manual-of-investigative-operations-and-guidelines-miog-part-02-01-of-06/view.
63. FBI Employment Agreement, Including Provisions and Prohibited Disclosures, FD-291, available at https://www.fbi.gov/file-repository/fd-291.pdf/view.
64. 44 U.S.C. 3101, "Records Management by Agency Heads, General Duties"; 5 U.S.C. 552a, "Records Maintained on Individuals"; 28 U.S.C. 1732, "Record Made in Regular Course of Business."
65. *Politico* Staff, "Full Text: James Comey Testimony Transcript on Trump and Russia," *Politico*, p. 16, June 8, 2017; https://www.politico.com/story/2017/06/08/full-text-james-comey-trump-russia-testimony-239295.
66. 18 U.S.C. 641, "Public Money, Property or Records."
67. 5 C.F.R. 2635.703, "Use of Nonpublic Information"; 29 C.F.R. 71.14, "Use of NonPublic Information."
68. Federal Bureau of Investigation, Records Management Division, 0792PG, p. 1, June 4, 2015.
69. FBI Employment Agreement, Including Provisions and Prohibited Disclosures, FD-291, available at https://www.fbi.gov/file-repository/fd-291.pdf/view.
70. 18 U.S.C. 793, "Gathering, Transmitting or Losing Defense Information."
71. *Politico* Staff, "Full Text: James Comey Testimony Transcript on Trump and Russia."
72. Letter from Charles E. Grassley, chairman of Committee on the Judiciary Committee, to Rod J. Rosenstein, deputy attorney general, January 3, 2018, available at https://www.judiciary.senate.gov/imo/media/doc/2018-01-03%20CEG%20to%20DOJ%20(Classification%20of%20Comey%20Memos).pdf.
73. Ibid.
74. Brooke Singman, "Comey Memos Reportedly Had Classified Info," Fox News, July 10, 2017.
75. 18 U.S.C. 1924, "Unauthorized Removal and Retention of Classified Documents or Material."
76. 18 U.S.C. 793(f), "Gathering, Transmitting or Losing Defense Information."
77. Catherine Herridge, Pamela K. Browne, and Cyd Upson, "Comey Memos Shared More Broadly Than Previously Thought," Fox News, April 25, 2018.
78. Transcript of "James Comey on Clinton Probe, Russia Investigation," Fox News, "Special Report With Bret Baier." April 26, 2018, http://www.foxnews.com/transcript/2018/04/26/james-comey-on-clinton-probe-russia-investigation.html.
79. Catherine Herridge, Pamela K. Browne, and Cyd Upson, "Comey Memos Shared More Broadly Than Previously Thought."
80. Ibid.; Sean Davis, "Comey 'Friend' Who Leaked FBI Memos Now Claims to Be His Attorney," *Federalist*, January 23, 2018.
81. Transcript of "James Comey on Clinton Probe, Russia Investigation," Fox News, "Special Report With Bret Baier." April 26, 2018; http://www.foxnews.com/transcript/2018/04/26/james-comey-on-clinton-probe-russia-investigation.html.

82. Michael D. Shear and Nicholas Fandos, "GOP Push on Comey Files May Have Backfired," *New York Times*, April 21, 2018.
83. United States Code, 18 U.S.C. 1001, "Statements or Entries Generally."
84. Hearing Before the Committee on the Judiciary, "Oversight of the Federal Bureau of Investigation," Testimony of James B. Comey, September 28, 2016, available at https://judiciary.house.gov/wp-content/uploads/2016/09/114-91_22125.pdf.
85. Senate Judiciary Committee letter to FBI Director Christopher Wray, August 30, 2017, available at https://www.judiciary.senate.gov/imo/media/doc/2017-08-30%20CEG%20+%20LG%20to%20FBI%20(Comey%20Statement).pdf.
86. Statement by FBI Director James B. Comey on the Investigation of Secretary Hillary Clinton's Use of a Personal Email System, FBI National Press Office, July 5, 2016.
87. Brooke Singman, Alex Pappas, and Jake Gibson, "More Than 50,000 Texts Exchanged Between FBI Officials Strzok and Page, Sessions Says," Fox News, January 22, 2018.
88. *Washington Post* Staff, "Read Full Testimony of FBI Director James Comey in Which He Discusses Clinton Email Investigation."
89. CNN Staff, "Read: Former FBI Deputy Director Andrew McCabe's Statement on His Firing," CNN, March 17, 2018, available at https://www.cnn.com/2018/03/16/politics/mccabe-fired-statement-fbi-deputy-director/index.html.
90. Letter from Sen. Charles E. Grassley and Sen. Lindsey O. Graham, Judiciary Committee, to Michael Horowitz, Inspector General, Department of Justice, February 28, 2018, available at https://www.grassley.senate.gov/sites/default/files/judiciary/upload/2018-02-28%20CEG%20LG%20to%20DOJ%20OIG%20%28referral%29.pdf
91. Byron Tau and Aruna Viswanatha, "Justice Department Watchdog Probes Comey Memos Over Classified Information," *Wall Street Journal*, April 20, 2018.
92. Interview with Joseph diGenova, former U.S. attorney for the District of Columbia and former independent counsel, January 26, 2018.

CHAPTER 11: THE ILLEGITIMATE APPOINTMENT OF ROBERT MUELLER

1. Associated Press, "Law Is 'Horrible', Says Darrow, 79," *New York Times*, April 19, 1936.
2. Ibid.
3. Code of Federal Regulations, 28 C.F.R. 600.1, "Grounds for Appointing A Special Counsel."
4. Code of Federal Regulations, 28 C.F.R. 600.4, "Jurisdiction."
5. *Washington Post* Staff, "Full Transcript: FBI Director James Comey Testifies on Russian Interference in 2016 Election," *Washington Post*, March 20, 2017.
6. Andrew C. McCarthy, "Rosenstein Fails to Defend His Failure to Limit Mueller's Investigation," *National Review*, August 7, 2017.

7. Order Issued by Acting Attorney General, Rod Rosenstein (Order No. 3915-2017), "Appointment of Special Counsel to Investigate Russian Interference With the 2016 Presidential Election and Related Matters," Office of the Deputy Attorney General, May 17, 2017.
8. Michael B. Mukasey, "The Memo and the Mueller Probe," *Wall Street Journal*, February 4, 2018.
9. Ibid.
10. *PBS Newshour,* "Justice Department Defends Russia Probe from GOP Claims of FBI Political Bias," PBS, December 13, 2017.
11. Order Issued by Acting Attorney General Rod Rosenstein (Order No. 3915-2017), "Appointment of Special Counsel to Investigate Russian Interference With the 2016 Presidential Election and Related Matters," Office of the Deputy Attorney General, May 17, 2017.
12. Ibid.
13. Code of Federal Regulations, 28 C.F.R. 600.4, "Jurisdiction."
14. Memorandum from Rod J. Rosenstein, Acting Attorney General, to Robert S. Mueller, Special Counsel, U.S. Department of Justice, August 2, 2017; Attachment C, Government's Response in Opposition to Motion to Dismiss, United States District Court for the District of Columbia (No. 17-cr-201-1 ABJ), April 2, 2018.
15. Ibid.
16. *Weeks v. United States,* 232 U.S. 383 (1914); *Mapp v. Ohio*, 368 U.S. 643 (1961).
17. Memorandum from Rod J. Rosenstein, Acting Attorney General, to Robert S. Mueller, Special Counsel, U.S. Department of Justice, August 2, 2017; Attachment C, Government's Response in Opposition to Motion to Dismiss, United States District Court for The District of Columbia (No. 17-cr-201-1 ABJ), April 2, 2018.
18. Interview by Gregg Jarrett of Alan Dershowitz, *Lou Dobbs Tonight,* Fox Business Network, April 6, 2018; *Hannity*, "Dershowitz and diGenova on Mueller Investigation," Fox News Channel, April 4, 2018, available at https://www.youtube.com/watch?v=P4u2YfDnAgc.
19. Government's Response in Opposition to Motion to Dismiss, United States District Court for the District of Columbia (No. 17-cr-201-1 ABJ), p. 21, April 2, 2018.
20. Indictment, *U.S. v. Manafort*, U.S. District Court for the Eastern District of Virginia, February 22, 2018; https://www.justice.gov/file/1038391/download.
21. Mollie Hemingway, "Manafort Lawyers Claim Leaky Mueller Probe Has Provided No Evidence of Contacts With Russian Officials," *Federalist*, May 3, 2018.
22. Jake Gibson, "Federal Judge Accuses Mueller's Team of Lying, Trying to Target Trump: 'C'mon Man!," Fox News, May 4, 2018.
23. The Blaze, "Leaked Court Transcripts Reveal Showdown Between Judge and Mueller lawyer. It's a Beatdown," May 6, 2018, https://www.theblaze.com/news/2018/05/06/leaked-court-transcripts-reveal-showdown-between-Judge-and-mueller-lawyer-its-a-total-beatdown. *U.S. v. Manafort*, "Defendant's Motion to Dismiss the Superseding Indictment," U.S. District Court for the District of Columbia, March, 14, 2018.

24. *Politico* Staff, "Full Text: James Comey Testimony Transcript on Trump and Russia," *Politico*, June 8, 2017.
25. Code of Federal Regulations, 28 C.F.R. 600.7, "Conduct and Accountability."
26. Code of Federal Regulations, 28 C.F.R. 45.2, "Disqualification Arising from Personal or Political Relationship."
27. Ibid.
28. Devlin Barrett, Adam Entous, Ellen Nakashima, and Sari Horowitz, "Special Counsel Is Investigating Trump for Possible Obstruction of Justice," *Washington Post*, June 14, 2017.
29. United States Code, 28 U.S.C. 528, "Disqualification of Officers and Employees of the Department of Justice."
30. Ronald D. Rotunda, "Alleged Conflicts of Interest Because of 'Appearance of Impropriety,'" Chapman University, Fowler School of Law, 2005, available at https://digitalcommons.chapman.edu/cgi/viewcontent.cgi?article=1001&context=law_articles.
31. Carl M. Cannon, "Comey, Mueller Bungled Big Anthrax Case Together," *Orange County Register*, May 21, 2017.
32. Ibid.
33. Peter Holley, "Brothers in Arms: The Long Friendship Between Mueller and Comey," *Washington Post*, May 17, 2017.
34. Garrett M. Graff, "Forged Under Fire—Bob Mueller and James Comey's Unusual Friendship," *Washingtonian*, May 30, 2013.
35. William G. Otis, "Robert Mueller Should Recuse Himself from Russia Investigation," *USA Today*, June 14, 2017; Glenn Harland Reynolds, "We Need a Robert Mueller Resignation or a Second Special Counsel," *USA Today*, June 19, 2017.
36. Jonathan Turley, "The Special Counsel Investigation Needs Attorneys Without Conflicts," *The Hill*, December 8, 2017.
37. Sidney Powell, *Licensed to Lie: Exposing Corruption in The Department of Justice* (Dallas: Brown Books, 2014).
38. Sidney Powell, "Political Prosecution: Mueller's Hit Squad Covered for Clinton and Persecutes Trump Associates," *Daily Caller*, December 6, 2017.
39. Byron York, "Is Robert Mueller Conflicted in Trump Probe?," *Washington Examiner*, June 11, 2017.
40. Code of Federal Regulations, 28 C.F.R., "Grounds for Appointing a Special Counsel."
41. Carol D. Leonnig and Robert Costa, "Mueller Told Trump's Attorneys the President Remains Under Investigation but Is Not Currently a Criminal Target," *Washington Post*, April 3, 2018.
42. Memorandum for the Attorney General, "Restoring Public Confidence in the FBI," Rod Rosenstein, deputy attorney general, May 9, 2017.
43. Sari Horwitz, Karon Demirjian, and Elise Viebeck, "Rosenstein Defends His Controversial Memo Used to Justify Trump's Firing of Comey," *Washington Post*, May 19, 2017.
44. Code of Federal Regulations, 28 C.F.R. 600.7, "Conduct and Accountable."
45. Ibid.

46. Aruna Viswanatha and Del Quentin Wilber, "Special Counsel's Office Interviewed Deputy Attorney General Rod Rosenstein," *Wall Street Journal*, September 19, 2017.
47. Sari Horwitz, Karoun Demirjian, and Elise Viebeck, "Rosenstein Defends His Controversial Memo Used to Justify Trump's Firing of Comey," *Washington Post*, May 19, 2017.
48. Rules of Professional Conduct, Rule 1.7—Conflict of Interest, "General Rule," available at https://www.dcbar.org/bar-resources/legal-ethics/amended-rules/rule1-07.cfm.
49. Tweet of Donald J. Trump, Twitter, June 16, 2017, available at https://twitter.com/realdonaldtrump/status/875701471999864833?lang=en.
50. Jonathan Turley, "Do Rosenstein and Mueller Have Conflicts of Interest in the Trump Investigation?," JonathanTurley.com, June 19, 2017, available at https://jonathanturley.org/2017/06/19/do-rosenstein-and-mueller-have-conflicts-of-interest-in-the-trump-investigation/.
51. Jonathan Turley, "It's High Time Rod Rosenstein Recuse Himself," *The Hill*, August 7, 2017.
52. Jack Goldsmith, "Why Hasn't Rod Rosenstein Recused Himself from the Mueller Investigation?," *Lawfare*, January 5, 2018.
53. Carrie Johnson, "Deputy Attorney General Rosenstein Supervises Mueller Probe but He's Also a Witness," National Public Radio, November 11, 2017.
54. Aruna Viswanatha and Del Quentin Wilber, "Special Counsel's Office Interviewed Deputy Attorney General Rod Rosenstein," *Wall Street Journal*, September 19, 2017.
55. *Hannity*, "Dershowitz and diGenova on Mueller Investigation," Fox News Channel, April 4, 2018, available at https://www.youtube.com/watch?v=P4u2YfDnAgc.
56. Brooke Singman, "Special Counsel Mueller's Team Has Only One Known Republican," Fox News, February 23, 2018.
57. Robert S. Mueller III, United States Department of Justice, available at https://www.justice.gov/criminal/history/assistant-attorneys-general/robert-s-mueller.
58. CNN Wire Staff, "Obama Requests 2 More Years for FBI Chief," CNN, May 13, 2011.
59. David Sivak, "Exclusive: Not a Single Lawyer Known to Work for Mueller Is a Republican," *Daily Caller*, February 21, 2018.
60. Brooke Singman, "Special Counsel Mueller's Team Has Only One Known Republican."
61. Ibid.
62. Peter Nicholas, Aruna Viswanatha, and Erica Orden, "Trump's Allies Urge Harder Line as Mueller Probe Heats Up," *Wall Street Journal*, December 8, 2017; David Shortell, "Mueller Attorney Praised Yates as DOJ Official Emails Show," CNN, December 5, 2017.
63. Brooke Singman, "More Clinton Ties on Mueller Team: One Deputy Attended Clinton Party, Another Rep'd Top Aide," Fox News, December 8, 2017.

64. Sidney Powell, "Meet the Very Shady Prosecutor Robert Mueller Has Hired for the Russia Investigation," *Daily Caller*, November 20, 2017.
65. Sidney Powell, "Political Prosecution: Mueller's Hit Squad Covered for Clinton and Persecutes Trump Associates," *Daily Caller*, December 6, 2017; Sidney Powell, "In Andrew Weissmann, the DOJ Makes a Stunningly Bad Choice for Crucial Role," *Observer*, January 12, 2015.
66. Sidney Powell, "Judging by Mueller's Staffing Choices, He May Not Be Very Interested in Justice," *The Hill*, October 19, 2017
67. Brooke Singman, "Top Mueller Investigator's Democratic Ties Raise New Bias Questions," Fox News, December 7, 2017.
68. Brooke Singman, "More Clinton Ties on Mueller Team: One Deputy Attended Clinton Party, Another Rep'd Top Aide," Fox News, December 8, 2017.
69. Josh Gerstein, "Another Prosecutor Joins Trump-Russia Probe, *Politico*, September 15, 2017.
70. Ibid.
71. Josh Delk, "GOP Senator: Mueller 'Needs to Clean House of Partisans,'" *The Hill*, December 16, 2017; Ken Dilanian, "Republicans Step Up Attacks on Special Counsel Robert Mueller," NBC News, December 13, 2017.

EPILOGUE

1. Franklin Foer, "Putin's Puppet: If the Russian President Could Design a Candidate to Undermine American Interests—and Advance His Own—He'd Look a Lot Like Donald Trump," *Slate*, July 4, 2016; Jeffrey Goldberg, "It's Official: Hillary Clinton Is Running Against Vladimir Putin," *Atlantic*, July 21, 2016; David Remnick, "Trump and Putin: A Love Story," *New Yorker*, August 3, 2016.
2. Abigail Tracy, "Is Donald Trump a Manchurian Candidate?," *Vanity Fair*, November 1, 2016; Aiko Stevenson, "President Trump: The Manchurian Candidate?" *Huffington Post*, January 18, 2017; Ross Douthat, "The 'Manchurian' President?," *New York Times*, May 31, 2017.
3. Colby Hall, "George Stephanopoulos Challenges Jay Sekulow: 'Cooperation Is Collusion!,'" Mediaite, October 31, 2017, available at: https://www.mediaite.com/tv/george-stephanopoulos-to-jay-sekulow-cooperation-is-collusion.
4. Rebecca Savransky: "Watergate Reporter Says Current White House 'Potentially More Dangerous Situation'," *The Hill*, May 14, 2017.
5. Tim Haines, "Dan Rather: "'Hurricane Vladimir' Is 'Approaching Category Four' for Trump Presidency," RealClearPolitics, August 31, 2017.
6. Aiden McLaulin, "Jake Tapper Goes Cronkite on Don Jr. Emails: 'Evidence of Willingness to Commit Collusion'," Mediaite, July 11, 2017.
7. Dan Gainor, "Impeach Trump? Liberal Media Profiteering from Anti-Trump Clickbait," Fox News, May 30, 2017.
8. Rachel Maddow transcripts, *The Rachel Maddow Show*, MSNBC, July 11, 2017.
9. Tim Haines, "Mika Brzezinski: 'Noose' Tightening on Trump Family; Might Go to Jail for The Rest of Their Lives," RealClear Politics, December 5, 2017.
10. YouTube, "Chris Hayes on Trump-Russia Allegations: Why Is Everyone Act-

ing Guilty?," March 9, 2018, available at https://www.youtube.com/watch?v=aaHoR48ESRk.

11. Aidan McLaughlin, "MSNBC's Joy Reid: What If Trump Locked Himself in White House and Refused Arrest?" Mediaite, April 9, 2018.
12. Jennifer Harper, "Media Bias Continues: 90% of Trump Coverage in Last Three Months Has Been Negative, Study Says," *Washington Times*, December 12, 2017; Rich Noyes, "Bias: 1,000 Minutes For Trump/Russia 'Collusion' vs. 20 Seconds for Hillary/Russia Scandal," Media Research Center NewsBusters, October 25, 2017; Hadas Gold, "Study: 91 Percent of Coverage on Evening Newscasts Was Negative to Donald Trump," *Politico*, October 25, 2016; Art Swift, "Americans' Trust in Mass Media Sinks to New Low," Gallup, September 14, 2016.
13. Office Of Inspector General, "A Review of Various Actions By The Federal Bureau of Investigation and Department of Justice In Advance of The 2016 Election," U.S. Department of Justice, page 497, June 2018, available at: https://www.justice.gov/file/1071991/download.
14. Ibid.
15. Fox News Insider, "Rep. Gaetz: Time For Mueller To Show Collusion Evidence Or End Investigation," Fox News, December 20, 2017; Mary Kay Linge, "Trump Slams FBI's McCabe Over Planned Retirement," *New York Post*, December 23, 2017.
16. Daniel Chaitin, "Lisa Page Testified Investigators Had No Proof Of Collusion When Robert Mueller Was Appointed: Report," *Washington Examiner*, September 16, 2018; John Solomon, "Lisa Page Bombshell: FBI Couldn't Prove Trump-Russia Collusion Before Mueller Appointment," *The Hill*, September 16, 2018.
17. Catherine Herridge, "DOJ's Bruce Ohr Wrote Christopher Steele Was 'Very Concerned About Comey's Firing – Afraid They Will Be Exposed'," Fox News, August 17, 2018; John Solomon, "The Handwritten Notes Exposing What Fusion GPS Told DOJ About Trump," *The Hill*, August 9, 2018.
18. Ibid.
19. Rowan Scarborough, "Strzok Explains How Anti-Trump Documents Made Its Way from Clinton Camp to Eager FBI," *Washington Times*, July 15, 2018.
20. Catherine Herridge, "Nellie Ohr Invokes Spousal Privilege, Avoiding Questions on Steele Dossier," Fox News, October 19, 2018; Jeremy Herb, "Fusion GPS Contractor Nellie Ohr Doesn't Say Much at House Interview," CNN, October 19, 2018; Olivia Beavers, "Nellie Ohr Exercises Spousal Privilege in Meeting With House Panels," *The Hill*, October 19, 2018.
21. Karoun Demirjian, "House Russia Probe Witness Invokes Fifth Amendment as Trump Urges Firing of DOJ Official Connected to Dossier," *Washington Post*, October 16, 2018; Kelly Cohen, "GPS Founder Glenn Simpson Pleads the Fifth Before House Committees," *Washington Examiner*, October 16, 2018.
22. Lena Felton, "The Full Text of the Nunes Memo," *Atlantic*, February 2, 2018.
23. Judicial Watch, "FBI Records Show Dossier Author Deemed 'Not Suitable for Use' as Source, Show Several FBI Payments In 2016, August 3, 2018, available at: https://www.judicialwatch.org/press-room/press-releases/judicial-watch-

fbi-records-show-dossier-author-deemed-not-suitable-for-use-as-source-show-several-fbi-payments-in-2016/; Rowan Scarborough, "Christopher Steele Broke FBI Media Rules After Being 'Admonished,' Documents Show, *Washington Times*, August 4, 2018; Tom Winter, "FBI Releases Documents Showing Payments to Trump Dossier Author Steele," *NBC News*, August 3, 2018.

24. Catherine Herridge, "DOJ's Bruce Ohr Wrote Christopher Steele Was 'Very Concerned About Comey's Firing—Afraid They Will Be Exposed'," Fox News, August 17, 2018; John Solomon, "The Handwritten Notes Exposing What Fusion GPS Told DOJ About Trump," *The Hill*, August 9, 2018.
25. Judicial Watch, "FBI Records Show Dossier Author Deemed 'Not Suitable for Use' as Source, Show Several FBI Payments in 2016, August 3, 2018, available at: https://www.judicialwatch.org/press-room/press-releases/judicial-watch-fbi-records-show-dossier-author-deemed-not-suitable-for-use-as-source-show-several-fbi-payments-in-2016/; Rowan Scarborough, "Christopher Steele Broke FBI Media Rules After Being 'Admonished,' Documents Show," *Washington Times*, August 4, 2018.
26. Rowan Scarborough, "Christopher Steele Broke FBI Media Rules After Being 'Admonished,' Documents Show," *Washington Times*, August 4, 2018.
27. Catherine Herridge and Pamela Browne, "DOJ Releases FISA Docs that Formed Basis for Surveillance of Ex-Trump Adviser Carter Page," Fox News, July 22, 2018.
28. Gregg Re, "FBI Told FISA Court Steele Wasn't Source of Report Used to Justify Surveilling Trump Team, Documents Show," Fox News, July 22, 2018.
29. Judicial Watch, "Justice Department Discloses No FISA Court Hearings Held on Carter Page Warrants," August 31, 2018, available at: https://www.judicialwatch.org/press-room/press-releases/judicial-watch-justice-department-discloses-no-fisa-court-hearings-held-on-carter-page-warrants/; John Bowden, "Trump Blasts FBI, DOJ Over Report on Carter Page Surveillance Warrants," *The Hill*, September 1, 2018; Craig Bannister, "Warrants to Spy on Carter Page Obtained Without FISA Court Hearings, DOJ Filing Reveals," *CBS News*, August 31, 2018.
30. Catherine Herridge, "Top FBI Lawyer Baker Offers 'Explosive' Testimony on 'Abnormal' Handling of Russia Probe Into Trump Campaign: Lawmakers," Fox News, October 3, 2018; Karoun Demirjian, "Trump Allies, Offering No Specifics, Say Former FBI Official Gives 'Explosive' Testimony In Russia Probe," *Washington Post*, October 3, 2018.
31. Ibid.
32. John Solomon, "Collusion Bombshell: DNC Lawyers Met With FBI on Russia Allegations Before Surveillance Warrant," *The Hill*, October 3, 2018.
33. Adam Goldman and Michael S. Schmidt, "Rod Rosenstein Suggested Secretly Recording Trump and Discussed 25th Amendment," *New York Times*, September 21, 2018.
34. Nicholas Fandos and Adam Goldman, "Former Top FBI Lawyer Says Rosenstein Was Serious About Taping Trump," *New York Times*, October 10, 2018.
35. Catherin Herridge, "FBI Lawyer's Testimony At Odds With Rosenstein Denial On 'Wire' Report," Fox News, October 9, 2018.

36. Adam Goldman and Michael S. Schmidt, "Rod Rosenstein Suggested Secretly Recording Trump and Discussed 25th Amendment," *New York Times*, September 21, 2018.
37. John Solomon and Buck Sexton, "Hill.TV Interview Exclusive: Trump Eviscerates Sessions: 'I Don't Have an Attorney General'," *The Hill*, October 19, 2018.

ABOUT THE AUTHOR

Gregg Jarrett is a legal and political analyst for Fox News, and was an anchor at the network for fifteen years. Before joining Fox News, he was an anchor and correspondent for MSNBC and an anchor for Court TV. He is a former trial attorney. He lives in Stamford, Connecticut.